CONCRETE IN THE
MARINE ENVIRONMENT

Modern Concrete Technology Series

Edited by

Arnon Bentur
Building Research Station
TECHNION
Israel Institute of Technology
Faculty of Civil Engineering
Technion City
Haifa 32000
Israel

Sidney Mindess
Department of Civil Engineering
University of British Columbia
2324 Main Mall
Vancouver
British Columbia
V6T 1W5
Canada

Concrete in the Marine Environment

P. KUMAR MEHTA
Department of Civil Engineering,
University of California at Berkeley, USA

CRC Press
Taylor & Francis Group
Boca Raton London New York

CRC Press is an imprint of the
Taylor & Francis Group, an **informa** business
A TAYLOR & FRANCIS BOOK

CRC Press
Taylor & Francis Group
6000 Broken Sound Parkway NW, Suite 300
Boca Raton, FL 33487-2742

First issued in paperback 2019

© 1991 by Taylor & Francis Group, LLC
CRC Press is an imprint of Taylor & Francis Group, an Informa business

No claim to original U.S. Government works

ISBN-13: 978-1-85166-622-5 (hbk)
ISBN-13: 978-0-367-86632-7 (pbk)

British Library Cataloguing in Publication Data

Mehta, P. Kumar (Pkmehta Kumar)
  Concrete in the marine environment.
  1. Marine Structures. Concrete
  I. Title
  620.4162

  ISBN 1-85166-622-2

Library of Congress Cataloging-in-Publication Data

Mehta, P. K. (Povindar K.)
  Concrete in the marine environment/P. Kumar Mehta.
    p.   cm.
  Includes bibliographical references and index.
  ISBN 1-85166-622-2
    1. Concrete—Corrosion.   2. Sea-water corrosion.   I. Title.
  TA440.M385   1991
  620.1'3623—dc20   91-3079
                                                                CIP

Visit the Taylor & Francis Web site at
http://www.taylorandfrancis.com

and the CRC Press Web site at
http://www.crcpress.com

Contents

Foreword

Plain concrete is a brittle material, with low tensile strength and strain capacities. Nonetheless, with appropriate modifications to the material, and with appropriate design and construction methodologies, it is being used in increasingly sophisticated applications. Indeed, our understanding of cementitious systems has advanced to the point where these systems can often be 'tailored' for various applications where ordinary concretes are limited.

However, the results of the current research, which make these advances possible, are still either widely scattered in the journal literature, or mentioned only briefly in standard textbooks. They are thus often unavailable to the busy engineering professional. The purpose of the Modern Concrete Technology Series is to provide a series of volumes that each deal with a single topic of interest in some depth. Eventually, they will form a library of reference books covering all the major topics in modern concrete technology.

Recent advances in concrete technology have been obtained using the traditional materials science approach:

1. characterization of the microstructure;
2. relationships between the microstructure and engineering properties;
3. relationships between the microstructural development and the processing techniques; and
4. selection of materials and processing methods to achieve composites with the desired characteristics.

Accordingly, each book in the Series will cover both the fundamental scientific principles, and the practical applications. Topics will be discussed in terms of the basic principles governing the behaviour of the various cement composites, thus providing the reader with information valuable for engineering design and construction, as well as a proper background for assessing future developments.

The Series will be of interest to practitioners involved in modern concrete technology, and will also be of use to academics, researchers, graduate students, and senior undergraduate students.

Arnon Bentur
Sidney Mindess

Preface

In the construction world the 21st century will be known as the century of concrete in the oceans. There are a number of reasons for this prediction. Human population is expected to grow to more than six billion by the end of the 20th century. Improvement in living conditions around the world has not kept pace with this increase in the population. To improve the standard of living, the search for solutions has already provided a major impetus for the exploitation of coastal and undersea energy and mineral resources.

Many industrial materials, commonly used for structural purposes, do not show long-term durability in the marine environment. Portland cement concrete has proved to be an exception and is, therefore, increasingly used for the construction of concrete structures. During the fourth quarter of this century, already twenty oil and gas production platforms consisting of heavily reinforced and prestressed concrete elements have been built in the North Sea. Many sophisticated structures, such as superspan cantilever concrete bridges, undersea concrete tunnels, storm barriers, and man-made concrete islands are either under construction or under planning and design. Compared to structural steel elements, cost and durability considerations are tipping the scale in favor of reinforced and prestressed concrete elements because it is well accepted now that, from the standpoint of engineering behavior, both materials can be used to give satisfactory performance under loading conditions normally encountered in service.

Since concrete as a construction material is rapidly emerging as the most economical and durable solution to marine structures, it is important that tomorrow's structural designers and engineers acquire a basic knowledge of possible interactions between seawater and portland cement concrete. There is a large number of publications on the subject which are generally found in professional journals, trade magazines, and conference proceedings. This book is in response to a need for presentation of the available knowledge in a logical and comprehensive manner.

Chapter 1 of the book contains a description of the various types of marine structures, and some examples of sophisticated concrete structures that have either been recently built or under construction. In Chapter 2, a description of the marine environment is given including the chemistry of seawater, marine organisms, typical temperature and hydrostatic pressure conditions in the oceans, tidal action, storm waves, and impact from floating ice. Chapter 3 contains a review of the composition, microstructure, and engineering properties of normal portland cement concrete. In Chapter 4, several case histories of concrete deterioration as a result of long exposure to seawater are described. This chapter is followed by a comprehensive discussion, in Chapter 5, of the physical and chemical causes of deterioration of reinforced concrete in seawater. Discussed in particular are the phenomena of corrosion of reinforcing steel, expansion and cracking of concrete by freezing and thawing cycles, sulfate and alkali-silica attacks, crystallization pressure of salts, and micro-organism attack on concrete.

Chapter 6 contains comprehensive information on the selection of materials and mix proportions for concrete mixtures that may be required to show long-term durability in seawater. A new method of proportioning highly durable concrete, containing superplasticizing and pozzolanic admixtures, is described. Recommended construction practice (i.e., mixing, transport, placement and curing) is briefly discussed in Chapter 7. Since many old marine structures are in various states of deterioration, a rather detailed description of repair and rehabilitation of deteriorated concrete is presented in Chapter 8. In particular, procedures for evaluation of the condition of a structure before repairing, and guidelines for the selection of repair materials and methods are given.

The author hopes that this simple book will serve the need of practicing engineers involved in the design and construction of concrete sea structures. It is also hoped that the book is useful to civil engineering educators, who have the responsibility of training tomorrow's designers and builders of structures.

P. Kumar Mehta

Chapter 1

Marine Structures—An Introduction

CLASSIFICATION OF MARINE STRUCTURES

The term, 'marine structures', is normally applied to coastal berthing and mooring facilities, breakwaters and tidal barriers, dry docks and jetties, container terminals, and offshore floating docks and drilling platforms. Such a description of marine structures is based on their function, and is not useful for design purposes. A classification of marine structures, based on their most prominent design feature, is proposed by Buslov.[1] The author has grouped the wide variety of marine structures into five general categories: piled platforms, flexible bulkheads, gravity structures, rubble mounds, and floating structures.

Each group, shown in Fig. 1.1, includes several structural types which differ mainly in the way they resist the main loads. Again, there are several fabrication alternatives for each of these structural types, for example, cast-in-place, precast, cellular, and prestressed concrete, which represent another level of classification not shown in Fig. 1.1. According to Buslov,[1] the suggested classification identifies only the basic structural types. Various hybrids and combinations are also possible, such as rigid anchored walls, sheet-pile cells (filled-shell gravity structures retained by flexible bulkheads), composite steel–concrete sandwich structures,[2] etc.

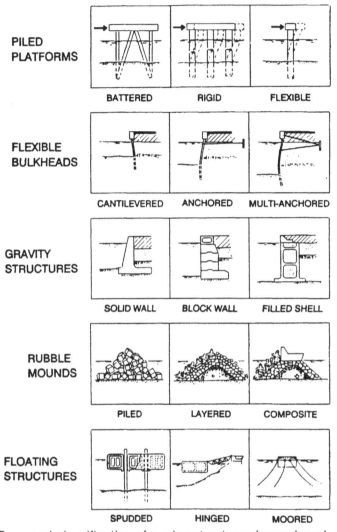

Fig. 1.1. Proposed classification of marine structures (reproduced, with permission, from Ref. 1).

CONCRETE IN THE MARINE ENVIRONMENT

Of the three basic structural materials, namely timber, steel, and concrete, reinforced concrete has become the most commonly used material for building marine structures. Reinforced portland cement concrete was invented approximately one hundred years ago, and it has become one of the most widely used industrial materials in the world. There are a number of reasons for this, such as the excellent resistance of concrete to water, the ease with which structural concrete elements can be formed into a variety of shapes and sizes, and the low cost and easy availability of concrete-making materials almost anywhere in the world.

Compared to other building materials, concrete also happens to show better resistance to the action of salt water; the vast numbers of wharfs, docks, bridge piers and beams, breakwaters, and subsea tunnels bear a clear testimony to its general acceptance as a suitable material of construction for structures exposed to the marine environment. In fact, during the last two decades, the use of concrete as the primary structural material has been extended to so many spectacular marine projects that Gerwick[3] foresees the world of concrete to be increasingly ocean-oriented. Others, including this author,[4] have also suggested that the twenty-first century will be the century of concrete in the oceans.

It is not difficult to justify the prediction that concrete will be increasingly used in both onshore and offshore sea structures. Since before recorded human history, oceans have been used for fishing, for commercial navigation, and for waste disposal. With increasing world population, which is approximately 5·5 (American) billion today, there has been a corresponding increase in coastal and offshore construction. More importantly, a rise in human expectations for a better standard of living has provided major impetus for the exploitation of undersea energy and mineral resources. Since the oceans cover twice as much surface as the combined area of the seven continents of the earth, already more than 25% of the world's hydrocarbons in the form of oil and gas are being extracted from coastal and offshore deposits. Beginning with Ekofisk, the world's first offshore concrete platform constructed in 1973 in the North Sea, there are now 20 oil and gas production platforms containing heavily reinforced and prestressed concrete elements.

Numerous marine construction projects involving complex structures, such as superspan bridges, undersea tunnels, breakwaters, and man-made islands, are under way. That concrete is now widely accepted as the preferred material of construction in the marine environment will be evident from some of the recently constructed or under-construction projects described below.

Offshore Concrete Platforms in the North Sea

According to Moksnes[5] the oil industry has historically been steel-oriented. Therefore, in 1972, it was a bold decision on the part of Phillips Petroleum Company to go for concrete as the principal structural material for the Ekofisk oilfield in the North Sea. Since then, a total of 20 concrete platforms containing approximately 2 million m³ of high-quality concrete have been built in the North Sea, in water depths ranging from 70 to 206 m (Table 1.1).

The Condeep type platform, pioneered by the Norwegian Contractors, is a concrete gravity base structure (GBS) with several shafts which support the deck (Fig. 1.2). The size of the caisson structure is governed by the needed crude oil storage facility, by the foundation area for structural stability during normal operation, and by the buoyancy as well as floating-

Table 1.1. North Sea Concrete Platforms.

Operator	Location	Design	Water depth (m)	Concrete vol. (m³)	Installation (year)
Phillips	Ekofisk	Doris	70	80 000	1973
Mobil	Beryl A	Condeep	118	52 000	1975
Shell	Brent B	Condeep	140	64 000	1975
Elf	Frigg CDP1	Doris	104	60 000	1975
Shell	Brent D	Condeep	140	68 000	1976
Elf	Frigg TP1	Sea Tank	104	49 000	1976
Elf	Frigg MP2	Doris	94	60 000	1976
Shell	Dunlin A	Andoc	153	90 000	1977
Elf	Frigg TCP2	Condeep	104	50 000	1977
Mobil	Statfjord A	Condeep	145	87 000	1977
Shell	Cormorant A	Sea Tank	149	120 000	1978
Chevron	Ninian Centr.	Doris	136	140 000	1978
Shell	Brent C	Sea Tank	141	105 000	1978
Mobil	Statfjord B	Condeep	145	140 000	1981
Mobil	Statfjord C	Condeep	145	130 000	1984
Statoil	Gullfaks A	Condeep	135	125 000	1986
Statoil	Gullfaks B	Condeep	141	100 000	1987
Norsk Hydro	Oseberg A	Condeep	109	120 000	1988
Statoil	Gullfaks C	Condeep	206	240 000	1989
Phillips	Ekofisk PB	Doris	75	105 000	1989

Reproduced, with permission, from Ref. 5.

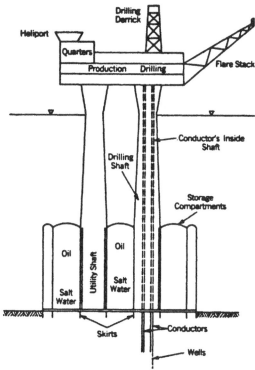

Fig. 1.2. Concrete gravity platform—Condeep type (courtesy: B. C. Gerwick).

stability requirements during the transportation and installation at the offshore location. The operations involved in the construction of a Condeep type platform are illustrated by the photographs in Fig. 1.3. A photograph of Gullfaks C, the world's largest offshore concrete platform, is shown in Fig. 1.4. Fabricated in a dry-dock at Hinna, near Stavanger, the 262 m tall concrete substructure consisting of 24 oil-storage cells and four shafts was mated with a 50 000-tonne steel deck, and in early 1989 the assembly was towed out to the installation site. A high-performance concrete mixture, 70–80 MPa compressive strength, was used for the construction of the caisson elements.

Concrete Island Drilling System

For drilling in the shallow waters of the Beaufort Sea, Alaska, a concrete island drilling system (CIDS) was pioneered in 1984 by a

(a)   (b)

(c)   (d)

(e)

Fig. 1.3. Sequence of operations in building a Condeep type platform (courtesy: Norwegian Contractors). (a) Construction of concrete base section in dry dock. (b) Slipforming of cell walls. (c) Slipforming of shafts. (d) Concrete structure ballasted into semisubmersed position to allow for mating with the deck. (e) Tow-out of the completed platform, Gullfaks A, on it's way to installation on the field in the North Sea.

Fig. 1.4. Gullfaks C—the world's largest offshore concrete platform (courtesy: J. Moksnes, Norwegian Contractors). Key data: water depth 216 m; production capacity 245 000 bbls/day; concrete volume 240 000 m³ (C65–70); reinforcement 70 000 tons; prestressing 3500 tons; deck weight during tow 49 500 tons.

consortium of American and Japanese companies. According to a report by Yee *et al.*,[6] the three basic components of the system, based on utilization of appropriate materials for varying environmental conditions, consist of a mud base, a concrete module, and a storage deck. The fully submerged mud base and the storage barges are not subjected to ice forces and are therefore constructed of structural steel. The $71 \times 71 \times 13$ m reinforced and prestressed concrete module, which is located in the splashing zone, is designed to resist bending, shear, and torsional forces from ice floes. Since the platform was constructed at a dry dock in Japan and then towed to the installation site in the Beaufort sea, structural strength and buoyancy were optimized by using two types of concrete mixtures for the module. A high-strength (55 MPa) normal-weight concrete mixture was used for the interior, but a high-strength (45 MPa) lightweight concrete mixture was used for the top and bottom slabs and for exterior walls. An expanded shale, sealed-pore type, lightweight aggregate was employed to produce the high-strength lightweight concrete.

Superspan Cantilever Concrete Bridges

Generally, prestressed concrete beam bridges are found to be more economical than steel bridges for spans ranging from 120 to 200 m. In Norway, 16 post-tensioned box-girder bridges with 150 m or more main spans were built during 1973–83, and the number of prestressed concrete bridges, with 100 to 150 m main spans built during the same period was considerably higher. In 1987, a 230 m long main-span bridge, the Norddalsfjorden Bridge, was constructed at a cost of only 4 million US dollars. The technology for building even longer-span bridges is already here. For such bridges, the dead load of the superstructure represents over 85% of the total load. Considerable reduction in dead load can be achieved by the use of high-strength lightweight concrete mixtures of the type described above. This is a highly effective method to reduce the overall dimensions, weight, and cost. A first-of-its-kind superspan bridge, the Helgeland Bridge, with a 390 m main span, was built in 1990. Other multispan concrete bridges, such as a bridge to cross the Straits of Gibraltar and a bridge to cross the Bering Strait, are being planned by structural designers.

According to T. Y. Lin,[7] a bridge from Alaska to Siberia across the Bering Strait would link six continents for travel, trade, and cultural exchange. Since this East–West linkage is envisaged to foster commerce and understanding between the people of the

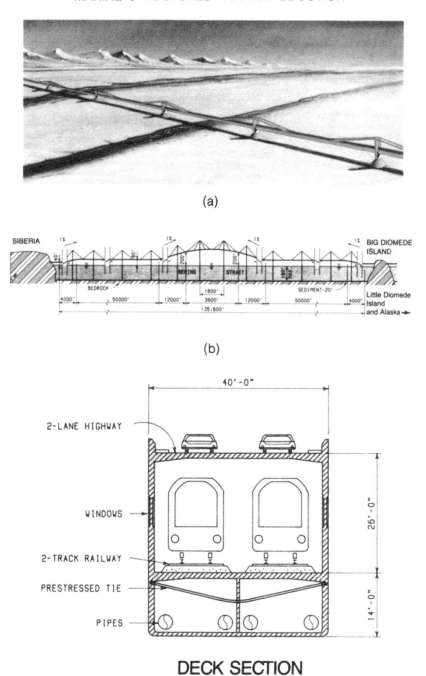

(a)

(b)

DECK SECTION

(c)

Fig. 1.5. (a) Intercontinental Peace Bridge. (b) Half elevation. (c) Deck section (courtesy: T. Y. Lin, Ref. 7).

United States and the Soviet Union, the proposed bridge is called the Intercontinental Peace Bridge. The design suggested by Lin consists of 220 spans, 200 m in length and approximately 23 m vertical clearance (Fig. 1.5). Each of the 220 concrete gravity piers is expected to require about 20 000 m³ normal-weight concrete; the superstructure will require 2·6 million m³ of lightweight concrete.

Undersea Tunnels

Undersea tunnels even though sited in soils well beneath the seafloor are surrounded by salt water, with an inside environment that supplies oxygen and heat.[3] Leakage ensures that there is a humid saline atmosphere in the interior as well as outside, with alternate wetting and drying due to ventilation and vehicle movement. Serious leakage has been reported from the Kanmon tunnels (between Honshu and Kyushu in Japan), the Hong Kong tunnels, the Al-Shindagha tunnel (under the estuary of the Dubai Creek), and the Suez tunnel. High-quality concrete is needed for tunnel lining because of the corrosive environment.

A photograph of the cross-section and longitudinal grade of the world's largest undersea tunnel, the 54 km long Seikan Railway Tunnel[8] in Japan, is shown in Fig. 1.6. Completed in 1988, 24 years after the beginning of its construction, the concrete-lined tunnel is about 100 m beneath the sea bottom at its midpoint and has both wide tracks for *Shinkansen* (Bullet) trains and narrow tracks for conventional trains. Since it provides an all-weather link between the overpopulated Honshu island and the underpopulated Hokkaido island, it is expected to play a vital role in the regional economic development of Hokkaido, which is rich in natural resources.

Eurotunnel, also called the Channel Tunnel, a 50 km undersea link between France and Great Britain, is reportedly the biggest infrastructure job to be privately financed in this century.[9] Designed for a service life of 120 years and scheduled for completion in 1992, this 9·2 billion US dollars project contains two 7·6 m i.d. rail tunnels, and a 4·8 m i.d. service tunnel, which are bored in chalk deposits 20 to 50 m below the seabed of the

Fig. 1.6. Cross-section and longitudinal grade of Seikan Tunnel, the world's largest undersea tunnel (courtesy: Japan Railway Constr. Public Corp.).

English Channel. Because of the poor rock quality, prefabricated reinforced concrete liners are used to line the rail tunnels. High-quality, relatively impermeable concrete mixtures were developed for the fabrication of these concrete liners[10] (see Chapter 6, Fig. 6.3).

In 1987, two subsea road tunnels, about 4 km long each, were built at Aalesund on the west coast of Norway. The tunnels descend to a depth of approximately 140 m below sea level. Because of the good rock quality, extensive concrete lining was not needed. However, where the rock quality was poor, shotcreting was done with a wet concrete mixture containing silica fume and steel fibers. The 'wet' shotcreting technique is also finding extensive application for repair of coastal structures (see Chapter 8).

Storm Barriers and Breakwaters

Storm surge barriers and breakwaters are frequently required to protect important coastal and offshore structures from high ocean waves. As described by Leenderste and Oud,[11] the Oosterschelde Storm Surge Barrier was built in 1986 in the southwestern part of the Netherlands. This area is situated at the delta of three rivers, the Rhine, the Maas, and the Schelde. The barrier is a part of the Delta Project which involves the closure of the main tidal estuaries and inlets in the southwestern part of the country. Not only did this shorten the country's coastline by hundreds of kilometers, but also the exclusion of saline water from a large area is expected to provide a significant improvement to freshwater management.

The 3000 m long storm surge barrier was built in three tidal channels. It consists of 65 prefabricated concrete piers, between which 62 sliding steel gates are installed (Fig. 1.7). With the gates in a raised position, the differences between the high and low tides behind the barrier are maintained at two-thirds of their original range, which is sufficient to preserve the natural environment of the Oosterschelde basin. When storms and dangerously high water levels are forecast, the gates can be closed in order to safeguard the population of the area from the ravages of the North Sea.

Fig. 1.7. Photograph of the Oosterschelde Storm Surge Barrier (courtesy: H. J. C. Oud, Ref. 11).

The Ekofisk concrete platform in the North Sea, built in 1973 in 70 m deep water at a distance of 170 km from the Norwegian coast, is protected with a breakwater to withstand up to $60\,t/m^2$ pressure from the 100-year high design wave. The breakwater consists of a perforated concrete wall (Fig. 1.8) which is designed to dampen the wave energy. Unreinforced elements of precast, high-strength (60–70 MPa compressive strength) and erosion resistant concrete, were used to construct the breakwater.

Recently it was discovered that, owing to the oil and gas extraction, the local seabed had sunk considerably more than expected. Consequently, the platform had to be jacked up and a special protective ring was necessary to protect the entire installation. A concrete ring, 108 m high and 140 m in diameter was constructed in a dry-dock in Rotterdam, the Netherlands, and installed in 1989.

Fig. 1.8. Ekofisk oil storage tanks and breakwater (reproduced with permission, from B. C. Gerwick, Jr, & E. Hognestad, *ASCE Environmental Design/Eng. Const.*, August, 1973).

SUMMARY

For the vast variety of marine structures in use in the twentieth century, concrete has emerged as the most economical and durable material of construction. During the last two decades, various technological hurdles have been successfully overcome to make reinforced and prestressed concrete a suitable building material for sophisticated structures, such as offshore drilling platforms, super-span bridges, and undersea tunnels. As the world[3] becomes increasingly ocean-oriented for energy and other resources, it is predicted that construction activity during the next century will be dominated by concrete sea structures.

REFERENCES

1. Buslov, V. M., New ACI Guide for design concrete marine structures, *Concrete International*, **12**(5) (1990), 41–6.
2. Gerwick, B. C. & Berner, D., Thermal and durability considerations for composite steel/concrete sandwich structures, *Performance of Concrete in Marine Environment*, ed. V. M. Malhotra, ACI SP-109, 1988, pp. 73–88.
3. Gerwick, B. C., Pressing needs and future opportunities in durability of concrete in marine environment, *Proceedings Gerwick Symposium on Durability of Concrete in Marine Environment*, ed. P. K. Mehta, Dept. of Civil Engineering, University of California at Berkeley, 1989, pp. 1–5.
4. Mehta, P. K., Durability of concrete in marine environment—an overview, *Proceedings Gerwick Symposium on Durability of Concrete in Marine Environment*, ed. P. K. Mehta, Dept. of Civil Engineering, University of California at Berkeley, 1989, pp. 20–7.
5. Moksnes, J., Oil and gas concrete platforms in the North Sea—reflections on two decades of experience, *Proceedings Gerwick Symposium on Durability of Concrete in Marine Environment*, ed. P. K. Mehta, Dept. of Civil Engineering, University of California at Berkeley, 1989, pp. 127–46.
6. Yee, A. A., Masuda, F. R., Kim, C. N., & Doi, D. A., Concrete module for the global marine concrete island drilling systems,

Proceedings FIP/CPCI Symposium on Concrete Sea Structures in Arctic Regions, Canadian Prestressed Concrete Institute, Ottowa, Canada, 1984, pp. 23–30.

7. Lin, T. Y., A concrete vision to re-unite the hemispheres, *Proceedings Gerwick Symposium on Durability of Concrete in Marine Environment,* ed. P. K. Mehta, Dept. of Civil Engineering, University of California at Berkeley, 1989, pp. 6–8.

8. Jenny, R. J., Twenty-three kilometers under the sea, *TR News,* No. 146, Jan.–Feb. (1990), 2–5.

9. *Engineering News Record,* Hard chalk slows tunnel bore, Nov. 3 (1988), 32–40.

10. Poitevin, P., Concrete for long service life: Channel Tunnel, *Dansk Beton,* No. 2 (1990), 71–6.

11. Leenderste, W. & Oud, H. J. C., The Dutch experience with construction and repair of marine structures, *Proceedings Gerwick Symposium on Durability of Concrete in Marine Environment,* ed. P. K. Mehta, Dept. of Civil Engineering, University of California at Berkeley, 1989, pp. 147–72.

Chapter 2
The Marine Environment

Seawaters throughout the world are characterized by both similarities and differences, which must be clearly understood before constructing concrete structures designed to last for hundreds of years. Also discussed in this chapter are general physical–chemical aspects of the marine environment, which are important from the standpoint of concrete durability.

CHEMICAL COMPOSITION OF SEAWATER

Most seawaters are similar with respect to the types and amounts of dissolved salts; the typical salt content is 3·5% by weight (35 g/liter or 35 parts per thousand parts), and the principal ions present are Na^+, Mg^{2+}, Cl^-, and $(SO_4)^{2-}$. The average composition of seawater is shown in Table 2.1.

In addition to dissolved salts, the presence of certain gases near the surface of seawater or in seawater also plays an important part in the chemical and electrochemical phenomena influencing concrete durability. For instance, oxygen (O_2) present in the atmospheric air and in seawater, either as entrapped air or dissolved oxygen, has an essential role in corrosion of steel in the marine environment regardless of whether the steel is exposed directly or embedded in concrete. Depending on local conditions, varying

17

Table 2.1. Average Composition of Seawater.

Ions	Concentration (g/liter)
Na^+	11·00
K^+	0·40
Mg^{2+}	1·33
Ca^{2+}	0·43
Cl^-	19·80
SO_4^{2-}	2·76

Reproduced, with permission, from Ref. 1.

concentrations of dissolved carbon dioxide (CO_2) and hydrogen sulfide (H_2S) may be found in seawater and may cause lowering of the pH from its normal value, 8·2–8·4, to 7 or even less. Acidic waters reduce the alkalinity and strength of concrete, and enhance the electrochemical corrosion of the embedded steel. Unusually large concentrations of free CO_2 are reported from seawaters in some sheltered bays and estuaries, when the seabed is covered with decaying organic matter. Typical sources of H_2S are marine organisms, as described next.

MARINE ORGANISMS

Marine growth involving barnacles and mollusks is frequently found on the surface of porous concrete whose alkalinity has been greatly reduced by leaching. Since marine growth is influenced by temperature, oxygen, pH, current, and light conditions, it is generally limited to about 20 m from the surface of the seawater and is less of a problem in cold climates.

Barnacles, sea urchins, and mollusks are known to secrete acids which can cause boreholes in concrete and pitting corrosion on the surface of embedded steel. As reported by Lea,[1] some mollusks produce ammonium carbonate, which is very damaging to concrete. According to Gerwick,[2] an aggressive variety of mollusks, which can bore into hard limestone aggregate in concrete, is found in the seawater of the Arabian/Persian Gulf area. H_2S-generating anaerobic bacteria are found in sediments containing oil. Known as, '*Theobacillus concretivorous*', they attack weak and permeable concrete, leading eventually to pitting corrosion of the embedded steel. As will be discussed in Chapter 5, the presence of aerobic or sulfur-oxidizing bacteria causes the conversion of H_2S to sulfuric acid, which is highly corrosive to both concrete and reinforcing steel.

According to Hoff,[3] marine growth may also be a problem because it can produce increased leg diameter and displaced volume which would result in increased hydrodynamic loading. The additional surface roughness provided by marine growth will increase the drag coefficient and will enhance the hydrodynamic loading, thus influencing structural stability (for instance, with large structures a 50 mm marine growth can easily add several thousand tons of dead weight). Marine growth also prevents adequate visual examination of concrete surfaces for other defects.

TEMPERATURE

The surface temperature of seawater varies widely from a low of −2°C (freezing point of seawater) in cold regions to a high of about 30°C in tropical areas. When the surface temperature is high, it shows a rapid drop with depth until the temperature reaches a steady-state value of about 2–5°C, at water depths of 100–1000 m.

In addition to its effect on the growth of marine organisms, the temperature of seawater determines the rate of chemical and electrochemical reactions in concrete. According to Idorn,[4] 'for concrete structures located in a warm climate, the heat may in itself be an aggravating factor because heat is a driving energy source which accelerates both the onset and the progress of the deteriorating mechanisms. The classical law which ties heat and the rate of chemical reactions together states that for each increase of ten degrees Celsius in temperature, the rate of chemical reactions is doubled. The fact that heat may have such a considerable impact on the rate of deterioration of reinforced concrete structures is only now beginning to be realized'.

It may be noted that air temperature shows a far greater degree of temperature extremes than seawater temperature. In tropical climates the daytime air temperature frequently reaches 40°C and in semi-enclosed areas, such as the Persian Gulf and the Arabian Sea, the air temperature may even reach 50°C. Under such conditions, the rate of evaporation is high resulting in high humidity. At the other extreme is the Arctic region where the air temperatures may reach as low as −40°C to −50°C. Such low temperatures may be below the transition temperature of some steels, leading to brittle fracture of the metal on impact. Tidal effects, when combined with the temperature difference between air and seawater, may expose some parts of a concrete sea structure to cycles of heating and cooling, freezing and thawing, and wetting and drying. The synergistic effect of these cycles can eventually destroy even the strongest of materials.

In order to focus the influence of climatic conditions on the durability of marine concrete structures, Fookes et al.[5] categorized the world climates into four types: cool (with freezing temperatures in the water), temperate (annual average temperature range 10 to 20°C, seldom freezing temperatures, moderate rainfall), hot and dry (such as the desert climate with summer temperatures exceeding 45°C and little rainfall), hot and wet (such as the tropical climate with annual average temperatures usually not exceeding 30°C). From case histories of the performance of reinforced concrete marine structures in each of the four climate zones, the authors concluded that, about 10 years after casting, the rate of deterioration was higher in the hot-dry and hot-wet climate zones when compared to the other zones. Three years after casting the differences were less significant, as shown in Table 2.2. Since the rates of deterioration of the individual segments of a concrete structure are also dependent on their location with respect to the tidal zone, the authors presented a conceptual view of the degree of concrete deterioration with respect to both the tidal and climatic zones (Fig. 2.1). International experience with marine concrete structures is generally in conformity with this conceptual view.

THE MARINE ENVIRONMENT 21

Table 2.2. Effect of Climatic Environment on Reinforced Concrete in the Marine Environment.

| Type of defect | Crack type | Effect of climatic environment on condition class at: | | | | | | | |
| | | about 3 years after casting | | | | about 10 years after casting | | | |
		Cool with freezing	Temperate	Hot-dry	Hot-wet	Cool with freezing	Temperate	Hot-dry	Hot-wet
Crazing. Initial drying shrinkage. Laitance	Surface crack	1	1	2	1	1/2	1	2/3 patchy	1/2
Plastic shrinkage. Restrained drying shrinkage	Part way to reinforcement	1	1	2	1	2	2[a]	3	2
Plastic settlement. Unrestrained drying shrinkage	Across reinforcement	1	1	1/2	1/2	1/2 patchy	1	2/3	2[a]
Plastic settlement. Unrestrained drying shrinkage	Along reinforcement	1	1	2	2	2	2	4	2

[a] Some reinforcement unaffected.
Condition class: 1, uncracked or only minor cracks; 2, some cracks; 3, moderate cracking; 4, severe cracking; 5, severe cracking and spalling. For splash some go up one class. For immersed some go down one class.
Reproduced, with permission, from Ref. 5.

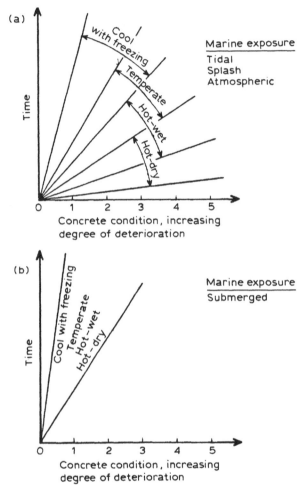

Fig. 2.1. Influence of different tidal and climatic zones on the performance of marine structures (Ref. 5).

HYDROSTATIC PRESSURE

The hydrostatic pressure of seawater on the submerged portion of a structure follows the simple relationship:

$$P = \rho h$$

where P is unit pressure, ρ is density of the fluid, and h is depth of water. Although the average density of seawater is $1026 \, \text{kg/m}^3$

$(64\,lb/ft^3)$, for practical purposes the hydrostatic pressure can be assumed as 1 tonne per square meter per meter of depth. The hydrostatic pressure acts as a driving force to push seawater through a permeable material. In the case of very porous concretes, capillary action augmented by the hydrostatic pressure may lead to migration of salt water to unsubmerged parts of the structure where the rapid rate of surface evaporation could cause salt crystallization pressures.

As Gerwick[2] has emphasized, it is important for the construction engineer to remember that full hydrostatic pressure can be exerted in even a relatively small hole, for example, an open prestressing duct or a duct left by removal of a slip-form climbing rod.

TIDAL ACTION

A tide comprises the gradual rise and fall of ocean water on a definite schedule twice a day. The gravity of the moon pulls the water nearest the moon slightly away from the solid part of the earth. At the same time, the moon pulls the solid earth slightly away from the water on the opposite side of the globe. In this way, the moon's gravity is constantly producing two bulges on the ocean waters during its daily journey around the earth. These bulges mark the position of high tide (highest water level).

At any coastal location the time interval between one high tide and the next is 12 hours and 25 minutes. As the earth turns, the tides rise and fall at each place on the ocean; a high tide is always followed by a low tide (lowest water level) after about 6 hours and 13 minutes. Tidal action takes place with remarkable regularity, i.e. from its low position seawater rises gradually for about six hours until it reaches the high tide mark and then it begins to fall for the next six hours until it reaches the low tide mark again. Thus, as a result of tidal action, a marine structure is exposed in the *tidal range* (between low and high tide levels) to twice-a-day cycles of wetting and drying, heating and cooling (due to differences between air and seawater temperatures), and possibly freezing and thawing (in cold climates).

The tidal range varies considerably from about 0·5 m in some locations to as much as 15 m in others. In funnel-shaped estuaries and bays, the ranges of the tide may be very high. For instance, in the Bay of Fundy on the east coast of Canada, the difference between high tide and low tide is sometimes more than 15 m (50 feet). The size, depth, and openness of an ocean also has an influence on the tidal range. The Atlantic and Pacific Oceans show regular flow and ebb, whereas the Mediterranean Sea has relatively little tidal action. The tidal range in the deep oceans is relatively small, usually less than 1 m; however, as one approaches the coast, the tidal range may increase to 4 or 5 m.

STORM WAVES

The forces exerted by ocean waves are enormous and are usually the primary design consideration affecting fixed structures. Waves are caused mainly by the action of wind on water; through friction the wind energy is transformed into wave energy. The typical

Fig. 2.2. Eddystone Lighthouse—1756. Reproduction of a painting made by the Danish artist Anton Melbye in 1846 (courtesy: Gunnar Idorn).

Table 2.3. Recent Concrete Breakwater Failures from Southern Europe, North Africa, and Turkey.

Antalya	Turkey	1971	600 m ruined
Arzew el Djedid	Algeria	1979/80	severe damage
Tripoli	Libya	1973/77	1981 severe damage
		1976/80	much during construction
Bilbao	Spain	1976	severe damage
Sines	Portugal	1978	severe damage
San Ciprian	Spain	1979	1980 severe damage

Reproduced, with permission, from Ref. 4.

fury of storm waves at the western outlet of the English Channel is well illustrated by a painting of the Eddystone Lighthouse (Fig. 2.2). At Wick in Scotland, one 800-ton anchored concrete block was lifted over the breakdown structure in a gale, and some years later a breakwater unit weighing 2600 tons was displaced in another gale.[4] Some recent concrete breakwater failures in Southern Europe, North Africa, and Turkey, due to storm damage are listed in Table 2.3.

Typhoons, hurricanes, storms, landslides, and earthquakes are capable of generating exceedingly high and strong waves, because the total energy in a wave is proportional to the square of the wave height. According to Gerwick[2] the force of impact of breaking waves against the side of a structure may reach $30 \, t/m^2$. The part of a concrete structure that is subject to such intense wave action, called the *splashing zone,* is vulnerable to erosion from sand, gravel, floating ice, and other suspended solids in seawater.

FOG AND SPRAY

In summer, coastal fog is formed when warm air from land passes over a colder ocean. In winter, the colder air from land passes over the warmer and more humid environment of seawater. In both cases moisture condensation results in fog or low stratus clouds. Coastal fogs are frequently the carrier of fine droplets of seawater arising from the spray action.

Like wave action, spray is created by the action of wind on waves. Waves breaking against a structure or a coastal landmass can hurl tons of seawater into the air, where it is picked up by wind. Strong winds in stormy weather are known to carry seawater over long distances inland. Thus the durability problems arising from the corrosive action of seawater on concrete are not limited to offshore and coastal marine structures alone.

ICE IMPACT AND ICE ABRASION

In the Arctic, repeated impacts from ice floes can result in a considerable loss of the surface of concrete structures. As described by Hoff[6] the mechanism of ice abrasion loss is as follows:

> Pieces of ice, driven by wind and current, can possess significant kinetic energy, much of which is dissipated into the concrete when the ice collides with a concrete structure. Some of the energy is lost in the crushing of the ice. A large ice floe in open waters will, upon initial contact with a structure, both load the structure and begin to crush itself at the point of contact with the structure. As the driving forces of the floe continue to move it forward against the structure, the resistance of the structure continues to increase to a point where the floe experiences a local failure in the ice, usually in flexure, some distance from the initial point of contact with the structure. This momentarily eases the load on the structure. The original ice, now damaged by crushing and cracking, is shunted away by the moving floe and new, undamaged ice in the floe then collides with the structure. The characteristics of the ice and the floe, combined with the dynamic response of the structure, will establish a 'ratcheting' effect on the concrete surface, repeatedly loading and unloading it. With time, this repetitive loading behavior can destroy the effectiveness of the aggregate bond near the surface of the concrete and cause or propagate microcracks in the concrete matrix.

From field investigation of Finnish lighthouses Huovinen[7] reported 300 mm abrasion over 30 years in Helsinki and 15 to 50 mm abrasion over 22–24 years in the Gulf of Bothnia. The *in situ* compressive strength of concrete at water level was 53% to 58% of the strength above the water level. The author prepared abrasion plots from which the abrasion depth can be estimated as a function of the compressive strength of concrete and ice-sheet movement. A frost-resistant (air-entrained) concrete mixture with at least 70 MPa compressive strength is recommended for good resistance to ice abrasion.

CONCLUDING REMARKS

The marine environment is highly inhospitable for commonly used materials of construction, including reinforced concrete. Seawater contains corrosive ions and gases, and is a home to numerous marine organisms that are harmful to construction materials. Hydrostatic pressure and temperature extremes, capable of accelerating the process of deterioration in materials, are frequently encountered with coastal and offshore structures. Storm waves have destroyed even strong structures. In the north, the Arctic region is almost always covered with sea ice, while in the south the Antarctic contains huge icebergs. Thus, the hostile and highly complex ocean environment presents both a formidable challenge and a great opportunity to materials engineers.

REFERENCES

1. Lea, F. M., *The Chemistry of Cement and Concrete*, 3rd edn, Chemical Publishing Co., New York, 1971, p. 627.
2. Gerwick, B. C., *Construction of Offshore Structures*, Wiley–Interscience, 1986, pp. 16–24.
3. Hoff, G. C., The service record of concrete offshore structures in the North Sea, *Proceedings, International Conference on Concrete in the Marine Environment*, The Concrete Society, London, 1986, pp. 131–42.
4. Idorn, G. M., Marine concrete technology—viewed with Danish eyes, *Proceedings, Gerwick Symposium on Durability of Concrete in Marine Environment*, ed. P. K. Mehta, Dept. of Civil Engineering, University of California at Berkeley, 1989, p. 40.

5. Fookes, P. G., Simm, J. D. & Barr, J. M., Marine concrete performance in different climatic environments, *Proceedings, International Conference on Concrete in the Marine Environment*, The Concrete Society, London, 1986, pp. 115–30.
6. Hoff, G. C., High strength lightweight concrete for the Arctic, *Proceedings, International Conference on Concrete in the Marine Environment*, The Concrete Society, London, 1986, pp. 9–19.
7. Huovinen, S., Ice abrasion of concrete in Arctic sea structures, *ACI Materials J.*, **87**(3) (1990), 266–70.

Chapter 3
Composition, Microstructure, and Properties of Concrete

Although materials, mix proportions, and concreting practice for modern marine structures have considerably improved during the last two to three decades, it is useful to review briefly the composition and properties of ordinary structural concrete because, from the turn of the century until the 1940s, most marine construction was carried out with this type of concrete. Knowledge of the composition versus property relationships in ordinary concrete not only will help the understanding of the causes of deterioration of old sea-structures, but also will provide a background for recent advancements in concrete technology.

Ordinary structural concrete is generally produced by mixing together four components: portland cement, water, sand (fine aggretate, $75\,\mu m$ to $5\,mm$ particle size), and gravel or crushed rock (coarse aggregate, $>5\,mm$ size). For a variety of reasons, many concretes contain one or more admixture types as the fifth ingredient. The maximum size of coarse aggregate usually varies from 12 to $37\,mm$, depending on the size of the structural element, thickness of the cover on reinforcing steel, reinforcement spacing, and the type of available equipment for concrete placement.

Concrete mixtures are proportioned to give a desired consistency or slump when fresh, and a specified minimum compressive strength when hardened under a standard curing condition (28-d moist curing at 23°C). Ordinary reinforced concrete

Table 3.1. Typical Proportions of Materials in Ordinary Structural Concrete.

Material	3 000 psi (20 MPa) concrete		4 500 psi (30 MPa) concrete	
	(lb/yd³)	(kg/m³)	(lb/yd³)	(kg/m³)
Cement	450	270	600	360
Water	300	180	300	180
Fine aggregate	1 350	180	1 430	860
Coarse aggregate	1 940	1 160	1 740	1 040
Cement paste:				
% by mass		19		22
% by volume		27		29

mixtures, typically, are designed for 100–150 mm (4–6 in) slump and 20–30 MPa (3000–4500 psi) compressive strength. The cement content of ordinary concrete mixtures generally ranges from 270 to 350 kg/m³ (450–600 lb/yd³), and the water/cement ratio from 0·5 to 0·7 by mass. Typical proportions of materials for both 20 MPa (3000 psi) and 30 MPa (4500 psi) concrete mixtures are shown in Table 3.1.

In this chapter, a general description of concrete-making materials, chemistry of cement hydration, composition of the components that make up the concrete microstructure, and important engineering properties of concrete are briefly described. Procedures for selecting materials and mix proportions for concrete are described in Chapter 6.

CONCRETE-MAKING MATERIALS

In addition to water, hydraulic cements, aggregates, and admixtures are the principal concrete-making materials that are briefly described below.

Water

Impurities in the water used for concrete mixing, when present in excessive amount, may adversely affect the setting time, strength, and volume stability of the product. Generally, water containing less than 2000 ppm (parts per million) of total dissolved solids is satisfactory for concrete. Most municipal water contains less solid content than this.

Water of an unknown quality may be used provided it can be shown that it does not have an adverse influence on the setting time, strength development, and durability of concrete. The US Army Corps of Engineers specifications require that the 7- and 28-day compressive strengths of mortar cubes made with water from a new source should be at least 90% of the reference specimens made with distilled water. No other special tests for determining the quality of mixing water are generally needed. Examples of impurities requiring special attention are discussed below.

Seawater containing up to 35 000 ppm of dissolved salts is generally suitable for unreinforced concrete. Although the initial rate of strength development may not be affected or may even be slightly enhanced, usually the use of seawater as mixing water for concrete causes a moderate reduction in ultimate strength. These concretes may also show efflorescence. In reinforced concrete, the possible corrosive effect of seawater on the reinforcement must be considered.

Water containing algae when used as mixing water in concrete has the effect of entraining considerable amounts of air, with accompanying decrease in strength. In one instance, an increase in algae content from 0·09% to 0·23% caused 10·6% air entrainment with a 50% reduction in compressive strength.

Water containing mineral oil in small amount may not have any adverse effects, but if the mineral-oil content is high, it may retard the setting time and reduce the strength of concrete.

Hydraulic Cements

Hydraulic cements are defined as cements that harden by reacting with water to form a product which is water-resistant. For structural concrete, portland cement and its various modifications, such as those covered by the ASTM C 150 Standard Specification for Portland Cement, are the principal hydraulic cements in use today.

Portland cement is a fine powder which, by itself, is not a binder but develops its binding property as a result of hydration, i.e. by chemical reaction between the cement minerals and water. The cement is produced by pulverizing a clinker with a small amount of calcium sulfate (viz., 4 to 6% gypsum), the clinker being a multimineralic mixture of high-temperature reaction products between oxides of calcium and silicon, aluminum, and iron. A more detailed description of the portland cement manufacturing process can be found in standard textbooks on cement and concrete.[1,2]

The chemical compositions of the principal clinker minerals correspond approximately to C_3S,[†] C_2S, C_3A, and C_4AF, and their main characteristics are summarized in Table 3.2. From the knowledge of relative rates of reactivity of the individual minerals or compounds with water, heat of hydration, and products of hydration, it is possible to design portland cements with special properties, such as high early strength, low heat of hydration, high sulfate resistance, moderate heat of hydration, and moderate sulfate resistance. Also, durability of concrete to cycles of freezing and thawing may be enhanced by the use of a portland cement containing an interground air-entraining agent. The foregoing principles provide the basis for the production of eight *types of portland cements,* listed as follows, that are covered by the ASTM C 150 Standard Specification:

- Type I: For general use when the special properties specified for other types are not needed. No limits are imposed on any of the four principal minerals; however, their typical proportions are as follows: $C_3S = 45\%$ to 55%, $C_2S = 20\%$ to 30%, $C_3A = 8\%$ to 12%, and $C_4AF = 6\%$ to 10%. Approximately 85% particles are less than $75\,\mu m$ (No. 200 mesh), and the typical specific surface area is $350\,m^2/kg$ Blaine.
- Type IA: Air-entraining Type I portland cement.
- Type II: For general use when moderate sulfate resistance or

---

[†] The following abbreviations are commonly used in cement chemistry: $C = CaO$; $S = SiO_2$, $A = Al_2O_3$; $F = Fe_2O_3$; $\bar{S} = SO_3$; $H = H_2O$.

Table 3.2. Principal Compounds of Portland Cement and their Characteristics.

Approximate composition	$3CaO \cdot SiO_2$	$\beta 2CaO \cdot SiO_2$	$3CaO \cdot Al_2O_3$	$4CaO \cdot Al_2O_3 \cdot Fe_2O_3$
Abbreviated formula	C_3S	βC_2S	C_3A	C_4AF
Common name	Alite	Belite	—	Ferrite phase, Fss
Principal impurities	MgO, Al_2O_3, Fe_2O_3	MgO, Al_2O_3, Fe_2O_3	SiO_2, MgO, alkalis	SiO_2, MgO
Common crystalline form	Monoclinic	Monoclinic	Cubic, orthorhombic	Orthorhombic
Proportion of compounds present (%)				
Range	35–65	10–40	0–15	5–15
Average in ordinary cement	50	25	8	8
Rate of reaction with water	Medium	Slow	Fast	Medium
Contribution to strength				
Early age	Good	Poor	Good	Good
Ultimate	Good	Excellent	Medium	Medium
Heat of hydration	Medium	Low	High	Medium
Typical (cal/g)	120	60	320	100

moderate heat of hydration is desired. Since C_3A is primarily responsible for sulfate attack via ettringite formation, its content is limited to a maximum of 8%. In addition to a maximum limit of 8% on the C_3A content, ASTM C 150 limits the sum of C_3A and C_3S to a maximum of 58% when moderate heat of hydration is desired.

- Type IIA: Air-entraining Type II cement.
- Type III: For use when high early strength is desired. Typically, the C_3S content of this cement ranges between 50% and 65%, and the high early strength is in part due to the higher specific surface area, which is approximately 500 m²/kg Blaine.
- Type IIIA: Air-entraining Type III portland cement.
- Type IV: For use when low heat of hydration is desired. The ASTM C 150 Standard Specification achieves this by limiting the maximum permissible C_3S and C_3A contents to 35% and 7%, respectively. The cement is coarser than Type I portland

cement and its C_2S content typically ranges between 40% and 50%.

- Type V: For use when high sulfate-resistance is desired. The ASTM Standard Specification calls for a maximum limit of 5% on C_3A which applies when a sulfate expansion test is not required.

Although the eight types of portland cements covered by ASTM C 150 include three air-entraining cements, the concrete industry prefers cements without the air-entraining capability because the use of air-entraining admixtures during the concrete-mixing operation offers a better control on the desired amount of air and air-void distribution in the hardened product. Consequently, there is little demand for air-entraining portland cements. Similarly, low-heat cement (ASTM Type IV) is no longer manufactured in the United States because the use of mineral admixtures in combination with ASTM Type II cement provides a more cost-effective method for controlling the heat of hydration. It is estimated that the bulk (over 90%) of portland cement produced in the United States corresponds to ASTM Types I and II; the rest of the production consists of ASTM Type III, Type V, and special products, such as oil-well cements, which are not covered by the ASTM Standards.

In addition to chemical requirements, hydraulic cements must comply with certain physical requirements such as fineness, soundness (freedom from cracking), time of set, and strength development rate. The principal physical requirements for Type I, II, III, and V portland cements, according to ASTM C 150, and the relevant test methods are summarized in Table 3.3. It may be noted here that the ASTM test methods and cement specifications are used mainly for the purpose of quality control of cements. They should not be used for predicting the properties of concrete, which are greatly influenced by the water–cement ratio, curing temperature, curing humidity, and cement-admixture interactions.

Blended portland cements and calcium aluminate cements belong to the category of *special hydraulic cements. Calcium aluminate cements,* which are no longer in use for structural purposes because of their strength retrogression tendency with aging, derive their

Table 3.3. Principal Physical Requirements and Essential Features of ASTM Test Methods for Portland Cements.

Requirement specified by ASTM C 150	Type I	Type II	Type III	Type V	Method of test
Fineness: minimum (m^2/kg)	280	280	None	280	ASTM Method C 204 covers determination of fineness of cements using Blaine Air Permeability Apparatus. Fineness is expressed in terms of specific surface of the cement
Soundness: maximum, autoclave expansion (%)	0·8	0·8	0·8	0·8	ASTM Method C 151 covers determination of soundness of cements by measuring expansion of neat cement paste prisms cured normally for 24 h and subsequently at 2 MPa (295 psi) steam pressure in an autoclave for 3 h
Time of setting					ASTM Method C 191 covers determination of setting time of cement pastes by Vicat apparatus. Initial setting time is obtained when the 1-mm needle is able to penetrate the 35-mm depth of a 40-mm-thick pat of the cement paste. Final setting time is obtained when a hollowed-out 5-mm needle does not sink visibly into the paste
Initial set minimum (min)	45	45	45	45	
Final set maximum (min)	375	375	375	375	
Compressive strength: minimum [MPa (psi)]					ASTM Method C 109 covers determination of compressive strength of mortar cubes composed of 1 part cement, 0·485 part water, and 2·75 parts graded standard sand by weight
1 day in moist air	None	None	12·4 (1 800)	None	
1 day moist air + 2 days water	12·4 (1 800)	10·3[a] (1 500)	24·1 (3 500)	8·3 (1 200)	
1 day moist air + 6 days water	19·3 (2 800)	17·2[a] (2 500)	None	15·2 (2 200)	
1 day moist air + 27 days water	None[b]	None[b]	None	20·7 (3 000)	

[a] The 3-day and the 7-day minimum compressive strength shall be 6·0 MPa (1000 psi) and 11·7 MPa (1700 psi), respectively, when the optional heat of hydration or the chemical limits on the sum of C_3S and C_3A are specified.

[b] When specifically requested, minimum 28-day strength values for Types I and II cements shall be 27·6 MPa (4000 psi).

cementing property from hydraulic calcium aluminates (CA, CA_2, and $C_{12}A_7$), rather than calcium silicates (C_3S and C_2S). They are useful for repair work, especially under cold weather or chemically aggressive conditions.[1,2] *Blended cements* consist essentially of an intimate blend of portland cement with a mineral admixture (pozzolanic or cementitious). Different types of mineral admixtures are described later; however, it may be noted here that two industrial by-products, namely fly ash and ground granulated blast-furnace slag, are among the most commonly used.

ASTM C 595, the Standard Specification for Blended Hydraulic Cements, covers five classes of blended cements, but commercial production in the United States is limited to Type IS and Type IP. According to the Specification, Type IS cement is an intimate and uniform blend of portland cement and finely pulverized granulated blast-furnace slag, in which the slag constituent is between 25% and 70% of the weight of portland blast-furnace slag cement (industrial cements typically contain 30% to 50% slag). Blast-furnace slag is a nonmetallic by-product of the iron industry, and consists essentially of silicates and aluminosilicates of calcium and magnesium. Granulated slag is produced by rapidly chilling the molten slag, and consists essentially of noncrystalline or glassy phases which possess hydraulic property. Type IP cement is generally a blend of portland cement with a fine pozzolan, in which the pozzolanic constituent can vary from 15% to 40%. Industrial Type IP cements typically contain 20% to 25% pozzolan.

Although cost and energy savings are among the main motivating factors for producing blended cements, the technical benefits associated with their use, such as the resistance to thermal cracking and chemical attacks, are being increasingly exploited by the concrete industry. In some countries of Europe and Asia, more than 25% of the total cement production consists of blended portland cements.

Hydration of Portland Cement

As stated earlier, anhydrous cement is not a binder; it acts as a binder only when mixed with water. This is because the chemical reactions between cement minerals and water form hydration products, which possess setting and hardening properties. Since portland cement is composed of a heterogeneous mixture of several minerals, the hydration process consists of simultaneously occurring reactions of the anhydrous minerals with water although the minerals do not hydrate at the same rate. The aluminates (C_3A and C_4AF) are known to hydrate at a much faster rate, and therefore the stiffening (loss of consistency) and setting (solidification) of portland cement pastes is generally attributed to the aluminate minerals. The hardening of cement paste (strength development) is attributable mainly to the silicate minerals, which make up about 75% of ordinary portland cement. The hydration reactions of the aluminate minerals and the silicate minerals in portland cement are briefly described below.

Hydration of the Aluminates

C_3A reacts with water rapidly to form crystalline products such as C_3AH_6, C_4AH_{19}, and C_2AH_8, with liberation of a large heat of hydration. Unless this rapid reaction is slowed down by some means, the cement paste would set and harden without allowing sufficient time for mixing, placement, and finishing of concrete. The retardation of the C_3A hydration reaction is generally accomplished by providing sulfate ions in water by intergrinding gypsum ($CaSO_4 \cdot 2H_2O$) with portland cement clinker. In any case, for practical purposes, what is important is not the hydration reaction of C_3A alone but the hydration of C_3A in the presence of gypsum. Also, for practical purposes, the hydration of the other aluminate mineral, namely C_4AF, will not be discussed here because the reaction products of C_4AF hydration in the presence of sulfate ions are similar in their crystal structure and properties to those formed by the hydration of C_3A. However, it may be noted that the C_4AF present in portland cement does not hydrate as rapidly as the C_3A.

When water is added to portland cement, gypsum and alkalis begin to ionize as rapidly as C_3A. However, within a few minutes the solubility of C_3A becomes depressed in the presence of sulfate and hydroxyl ions. Depending on the ionic concentration of aluminate and sulfate in the solution (or $A/\bar{S}$ ratio), the precipitating crystalline product is either the calcium aluminate trisulfate hydrate (also known by its mineralogical name, ettringite) or the calcium aluminate monosulfate hydrate. The relevant chemical reactions of hydration of C_3A in the presence of sulfate ions may be shown as follows:

$$[AlO_4]^- + 3[SO_4]^{2-} + 6[Ca]^{2+} + He \rightarrow C_6A\bar{S}_3H_{32} \quad \text{(ettringite)}$$

$$[AlO_4]^- + [SO_4]^{2-} + 4[Ca]^{2+} + He \rightarrow C_4A\bar{S}H_{18} \quad \text{(monosulfate)}$$

The chemical composition of ettringite and the monosulfate formed in the presence of iron may correspond to $C_6A(F)\bar{S}_3H_{32}$ and $C_4A(F)\bar{S}H_{18}$,[†] respectively. It may be noted that ettringite, in the form of short prismatic needles, is usually the first hydration product to crystallize from a freshly mixed portland cement paste because of low $A/\bar{S}$ ratio in the solution phase during the first hour of hydration. In normal portland cements containing 4% to 6% gypsum, the precipitation of ettringite is mainly responsible for the stiffening (slump loss) phenomenon, setting (solidification of the paste), and early strength. Later, during the hydration process, when the sulfate becomes depleted in the solution phase and the concentration of aluminate ions goes up, owing to renewed hydration of C_3A and C_4AF, ettringite will become unstable and will be gradually replaced by the monosulfate hydrate which, therefore, is the final product of the aluminate–sulfate interaction in portland cements containing more than 5% C_3A.

The monosulfate hydrate crystallizes in the form of thin hexagonal sheets that are similar in morphology and properties to some of the calcium aluminate hydrates. If for any reason, such as an inadequate amount of gypsum present, there should be an imbalance in the aluminate-to-sulfate ratio in the solution phase during the early period of cement hydration, and if as a result of this imbalance sheet-structure crystalline products are formed instead of ettringite, the cement paste will show abnormal stiffening and setting behavior.[1]

[†] In the cement chemistry literature the terms AF_t and AF_m are sometimes used to describe the products which may have a variable A/F ratio but are structurally similar to ettringite and monosulfate hydrate, respectively.

Incidentally, it is appropriate here to explain why ASTM Type V cement (less than 5% C_3A) is much more resistant to sulfate attack when compared to ASTM Type I cement (8% to 12% C_3A). In the former case, ettringite continues to persist as a stable product of hydration in the hardened cement paste. In the latter case, since the final product of hydration is the monosulfate hydrate, it would convert back to ettringite when exposed to sulfate water. In hardened concretes, formation of ettringite may be accompanied by expansion and cracking.

Hydration of the Silicates

The chemical composition and crystal structure of the calcium silicate hydrates, formed as a result of the hydration of C_3S and βC_2S, are similar. The chemical composition of the calcium silicate hydrates varies widely with the water-to-cement ratio, temperature, and age of hydration. It has become customary to refer to these hydrates collectively as 'C-S-H'—a notation that does not imply a fixed chemical composition. Because it is the crystal structure which determines the strength and other physical properties, the compositional differences are not of much significance from the standpoint of physical and mechanical characteristics. On complete hydration, the approximate chemical composition of the product corresponds to $C_3S_2H_3$; this composition is therefore used in stoichiometric equations:

$$2C_3S + 6H \rightarrow C_2S_2H_3 + 3CH$$

$$2C_2S + 4H \rightarrow C_3S_2H_3 + CH$$

The C_3S phase in portland cement is much more reactive than the C_2S. Although the C_3S phase is not as reactive as the C_3A, it begins to hydrate within the first hour after the addition of water to cement. The hydration products of C_3S, especially the C-S-H, contribute to the final set and the early strength of the cement paste. Since C-S-H is poorly crystalline, it has a high specific surface and shows excellent adhesive characteristics attributable to van der Waals forces of attraction. On the other hand, calcium hydroxide (CH) formed during early stages of hydration of portland cement pastes, is in the shape of massive hexagonal crystals, which have a low specific surface and poor bonding characteristics.

Physical Manifestations of Hydration Reactions

In concrete technology, from the standpoint of practical significance the physical manifestations of the hydration reactions of the aluminate and silicate phases are of considerable interest. As described next, the physical phenomena, such as stiffening, setting, and hardening of the cement paste, are different manifestations of the ongoing chemical processes.

Stiffening is associated with the slump loss phenomenon in fresh concrete and is due to a decrease in the consistency of the plastic cement paste. The free water in a cement paste determines its consistency. A portion of this free water is consumed by hydration reactions, some is lost by evaporation, and some is immobilized by physical adsorption on the surface of poorly crystalline hydration products such as C-S-H and ettringite. The gradual loss of free water from the system causes the paste to stiffen first, and set afterwards. The term *setting* signifies solidification of the cement paste. The beginning of solidification, as determined by the use of the Vicat Needle (ASTM C 191), is termed *initial set*. The initial set marks a point in time at which the cement paste has started to become unworkable but has not yet fully solidified. The placement, compaction, and finishing operations of concrete become very difficult beyond the point of 'initial set'; therefore, these operations must be completed before the initial set. The *final set* marks the point at which the cement paste has just completely solidified. The final set marks the beginning of the hardening (strength development) process, and it has an influence on construction scheduling.

At the time of final set, a concrete has little strength because the final set corresponds to the beginning of hydration of C_3S which is the main compound in portland cement. Once the C_3S hydration starts, the reaction is accelerated by the heat of hydration. The progressive replacement of the water-filled space in the cement paste with the hydration products results in a decrease of porosity

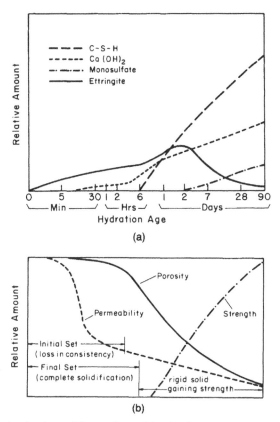

Fig. 3.1. (a) Typical rates of formation of hydration products in an ordinary portland cement paste; (b) influence of formation of hydration products on setting time, porosity, permeability, and strength of cement paste. ((a) Adapted from I. Soroka, *Portland Cement Paste and Concrete,* The Macmillan Press, 1979, p. 35, and reproduced with permission).

and permeability, and an increase in strength. A diagrammatic representation of the relationship between the chemical processes of hydration and the corresponding physical phenomena of stiffening, setting, and hardening is shown in Fig. 3.1.

Microstructure of Hydrated Portland Cement Pastes

The important engineering characteristics of hardened concrete, such as strength, durability, drying shrinkage, and creep, are greatly influenced by the properties of the hydrated cement paste

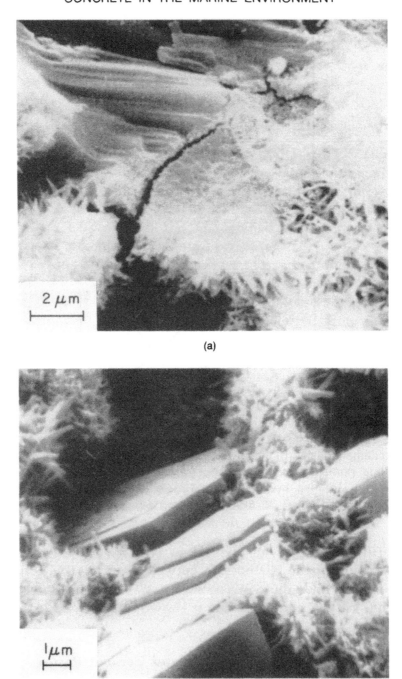

(a)

(b)

Fig. 3.2. Scanning electron micrographs of a fractured specimen of a 3-day-old portland cement paste.

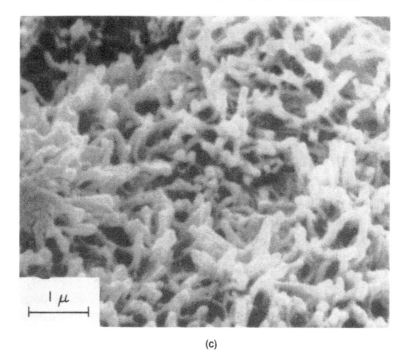

(c)

Fig. 3.2. (*Continued*)

which, in turn, depend upon its microstructural features, i.e. the type, amount, and distribution of solid phases and voids. Therefore, a brief description of the microstructure and structure–property relations of hydrated portland cement paste is given below.

Figure 3.2 shows scanning electron micrographs illustrating the typical morphology of portland cement hydration products, such as C-S-H and CH. Based on numerous scanning electron micrographic observations, a model showing the distribution of essential phases present in the microstructure of a well-hydrated portland cement paste is presented in Fig. 3.3.

In a well-hydrated system the aggregations of poorly crystalline C-S-H fibers constitute 50% to 60% of the total volume occupied by all solids, the massive stacks of hexagonal-plate CH crystals take up approximately 25% of the solids volume, and the remaining solids consist of needle-like crystals of ettringite, thin hexagonal sheets of monosulfate hydrate, and unhydrated clinker minerals. It has been stated earlier that the strength of hardened

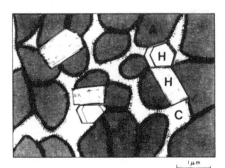

Fig. 3.3. Model of a well-hydrated portland cement paste. 'A' represents aggregation of poorly crystalline C-S-H particles which have at least one colloidal dimension (1 to 100 nm). Inter-particle spacing within an aggregation is 0·5 to 3·0 nm (average 1·5 nm). H represents hexagonal crystalline products such as CH, $C_4A\bar{S}H_{18}$, C_4AH_{19}. They form large crystals, typically 1 μm wide. C represents capillary cavities or voids which exist when the spaces originally occupied with water do not get completely filled with the hydration products of cement. The size of capillary voids ranges from 10 nm to 1 μm, but in well-hydrated, low water/cement ratio pastes, they are <100 nm.

cement paste is derived mainly from van der Waals forces of attraction present in C-S-H.

In addition to the solid phases, the hydrated cement paste contains several types of voids in the size range of 1 nm to 1 mm. The typical size of individual solid phases and void types is illustrated in Fig. 3.4. The interlayer void space within the C-S-H structure is reported to be of the order of 0·5 to 2·5 nm. This void size is too small to exercise any adverse influence on strength and permeability. Also, the small amount of water held by hydrogen bonding in this narrow space is difficult to remove under ordinary conditions of drying or loading, and thus should have little effect on drying shrinkage and creep. As discussed next, it is the capillary voids and air voids that influence many properties of the hydrated cement paste.

The total space in a freshly mixed, non-air-entrained, cement paste consists of unhydrated cement particles and water. In a hydrated cement paste, *capillary voids* represent the irregularly shaped, empty, or water-filled spaces that are not filled by the solid hydration products. The volume and the size of capillary

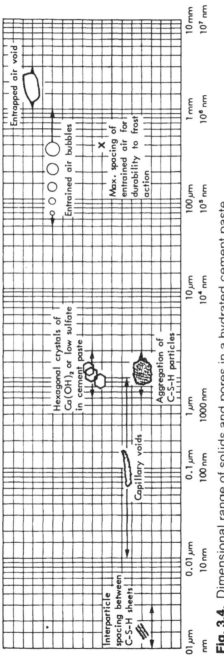

Fig. 3.4. Dimensional range of solids and pores in a hydrated cement paste.

voids are determined by the original distance between anhydrous cement particles in the fresh paste (i.e. the water-to-cement ratio) and the degree of cement hydration. In pastes with low water/cement ratios and high degrees of hydration, the volume of capillary voids may be under 10% of the hydrated cement paste volume, and their individual sizes may range from 5 to 50 nm. In pastes with high water/cement ratios, especially at early ages of hydration, the total volume of capillary voids may be as high as 40% to 50% of the hydrated cement paste, and their size may be as large as 3 to 5 μm. The capillary voids larger than 50 nm are considered to be detrimental to strength and impermeability, whereas voids smaller than 50 nm play a more important role in determining the drying shrinkage and creep characteristics of hydrated cement pastes. It is suggested that under the influence of attractive forces present in small capillary voids, up to six molecular layers of water can be physically held by hydrogen bonding. This water causes a disjoining pressure on the system, and its easy removal either by normal drying or by loading results in drying shrinkage or creep, respectively.

Furthermore, there are two types of air voids which, owing to their spherical shape, are readily distinguishable from the capillary voids. The *entrapped air voids,* sometimes as large as 3 mm, represent the small amount of air that usually gets entrapped during the mixing operation. The *entrained air voids,* ranging from 50 to 200 μm air bubbles, are purposely incorporated into a fresh cement paste by using an air-entraining type chemical admixture when it is intended to make the hardened product resistant to freezing and thawing cycles. Without the entrained air a saturated cement paste will expand on freezing because of the development of hydraulic pressure. Small bubbles of entrained air can get filled under pressure, and therefore are able to provide pressure relief when closely spaced within the hardened cement paste.[1]

Aggregates

Natural minerals form the most important class of aggregates for making portland cement concrete. Siliceous gravels or crushed rock are the common sources of coarse aggregate, while natural silica sand is predominantly used as the fine aggregate. Natural mineral aggregates are derived from rocks of many types and most rocks are themselves composed of several minerals. Thus, similar to portland cement particles, each particle of aggregate in concrete may be multimineralic.

The aggregate in concrete influences workability, unit weight, elastic modulus, and dimensional stability. To a large extent, these properties depend on the grading and the proportions of fine and coarse aggregate, as well as on aggregate porosity or bulk density. It seems therefore that with ordinary structural concrete, except for certain harmful ingredients when present, the chemical or mineralogical characteristics of the aggregate are less important than the physical characteristics, such as the shape and grading of particles, and the volume, size, and distribution of voids within a particle. The ASTM grading requirements for coarse aggregate and fine aggregate are shown in Table 3.4 and Table 3.5, respectively, and the ASTM C 33 limits for deleterious substances in concrete aggregates are shown in Table 3.6.

Admixtures

Many important properties of concrete, both freshly made and hardened, can be modified to advantage by the incorporation of certain materials into a concrete mixture. These materials, commonly known as admixtures, are now widely used by the concrete industry. Other books[1-4] may be consulted for a detailed description of various types of concrete admixtures; however, a brief review can be given here.

Admixtures can be categorized into two types: chemical and mineral. The chemicals used as admixtures can be further divided into two classes. Some chemicals begin to act on the cement–water system instantaneously by influencing the surface tension of water and by adsorption on the surface of cement particles; others ionize in water and will either accelerate or retard the solution of cement minerals during the early hydration. Mineral admixtures are finely ground, insoluble materials, derived either from natural sources or industrial by-products, which influence the properties

Table 3.4. Grading Requirements for Coarse Aggregates.

Size number	Nominal size (sieves with square openings)	Amounts finer than each laboratory sieve (square openings) (wt %)												
		100 mm (4 in)	90 mm ($3\frac{1}{2}$ in)	75 mm (3 in)	63 mm ($2\frac{1}{2}$ in)	50 mm (2 in)	37.5 mm ($1\frac{1}{2}$ in)	25.0 mm (1 in)	19.0 mm ($\frac{3}{4}$ in)	12.5 mm ($\frac{1}{2}$ in)	9.5 mm ($\frac{3}{8}$ in)	4.75 mm (No. 4)	2.36 mm (No. 8)	1.18 mm (No. 16)
1	90–37.5 mm ($3\frac{1}{2}$–$1\frac{1}{2}$ in)	100	90–100	—	25–60	—	0–15	—	0–5	—	—	—	—	—
2	63–37.5 mm ($2\frac{1}{2}$–$1\frac{1}{2}$ in)	—	—	100	90–100	35–70	0–15	—	0–5	—	—	—	—	—
3	50–25.0 mm (2–1 in)	—	—	—	100	90–100	35–70	0–15	—	0–5	—	—	—	—
357	50–4.75 mm (2 in–No. 4)	—	—	—	100	95–100	—	35–70	—	10–30	—	0–5	—	—
4	37.5–19.0 mm ($1\frac{1}{2}$–$\frac{3}{4}$ in)	—	—	—	—	100	90–100	20–55	0–15	—	0–5	—	—	—
467	37.5–4.75 mm ($1\frac{1}{2}$ in–No. 4)	—	—	—	—	100	95–100	—	35–70	—	10–30	0–5	—	—
5	25.0–12.5 mm (1–$\frac{1}{2}$ in)	—	—	—	—	—	100	90–100	20–55	0–10	0–5	—	—	—
56	25.0–9.5 mm (1–$\frac{3}{8}$ in)	—	—	—	—	—	100	90–100	40–85	10–40	0–15	0–5	—	—
57	25.0–4.75 mm (1 in–No. 4)	—	—	—	—	—	100	95–100	—	25–60	—	0–10	0–5	—
6	19.0–9.5 mm ($\frac{3}{4}$–$\frac{3}{8}$ in)	—	—	—	—	—	—	100	90–100	20–55	0–15	0–5	—	—
67	19.0–4.75 mm ($\frac{3}{4}$ in–No. 4)	—	—	—	—	—	—	100	90–100	—	20–55	0–10	0–5	—
7	12.5–4.75 mm ($\frac{1}{2}$ in–No. 4)	—	—	—	—	—	—	—	100	90–100	40–70	0–15	0–5	—
8	9.5–2.36 mm ($\frac{3}{8}$ in–No. 8)	—	—	—	—	—	—	—	—	100	85–100	10–30	0–10	0–5

Reprinted, with permission, from the Annual Book of ASTM Standards, Section 4, Vol. 04.02. Copyright, ASTM, 1916 Race Street, Philadelphia, PA 19103.

Table 3.5. Grading Requirements for Fine Aggregates.

Sieve (Specification E11)	Percentage passing
9·5-mm ($\frac{3}{8}$-in)	100
4·75-mm (No. 4)	95–100
2·36-mm (No. 8)	80–100
1·18-mm (No. 16)	50–85
600-μm (No. 30)	25–60
300-μm (No. 50)	10–30
150-μm (No. 100)	2–10

Reprinted, with permission, from the 1983 Annual Book of ASTM Standards, Section 4, Vol. 04.02. Copyright, ASTM, 1916 Race Street, Philadelphia, PA 19103.

Table 3.6. Limits for Deleterious Substances in Concrete Aggregates.

Substance	Possible harmful effects on concrete	Maximum permitted (wt %)	
		Fine aggregate	Coarse aggregate[a]
Material finer than 75-μm (No. 200) sieve			
Concrete subject to abrasion	Affect workability;	3[b] ⎫	1
All other concrete	increase water requirement	5[b] ⎭	
Clay lumps and friable particles	Affect workability and abrasion resistance	3	5
Coal and lignite			
Where surface appearance of concrete is important	Affect durability; cause staining	0·5 ⎫	0·5
All other concrete		1·0 ⎭	
Chert (less than 2·4 specific gravity)	Affect durability	—	5

[a] *ASTM C 33 limits for deleterious substances in coarse aggregate vary with the conditions of exposure and type of concrete structure. The values shown here are for outdoor structures exposed to moderate weather conditions.*

[b] *In the case of manufactured sand, if the material finer than 75-μm sieve consists of the dust of fracture, essentially free of clay or shale, these limits may be increased to 5% and 7%, respectively.*

Reprinted, with permission, from the 1983 Annual Book of ASTM Standards, Section 4, Vol. 04.02. Copyright, ASTM, 1916 Race Street, Philadelphia, PA 19103.

Table 3.7. Classification, Composition, and Particle Characteristics of Mineral Admixtures for Concrete.

Classification	Chemical and mineralogical composition	Particle characteristics
Cementitious and pozzolanic		
Granulated blast-furnace slag (cementitious)	Mostly silicate glass containing mainly calcium, magnesium, aluminum, and silica. Crystalline compounds of melilite group may be present in small quantity	Unprocessed material is of sand size and contains 10–15% moisture. Before use it is dried and ground to particles less than 45 μm (usually about 500 m^2/kg Blaine). Particles have rough texture
High-calcium fly ash (cementitious and pozzolanic)	Mostly silicate glass containing mainly calcium, magnesium, aluminum, and alkalis. The small quantity of crystalline matter present generally consists of quartz and C_3A; free lime and periclase may be present; $C\bar{S}$ and $C_4A_3\bar{S}$ may be present in the case of high-sulfur coals. Unburnt carbon is usually less than 2%	Powder corresponding to 10–15% particles larger than 45 μm (usually 300–400 m^2/kg Blaine). Most particles are solid spheres less than 20 μm in diameter. Particle surface is generally smooth but not as clean as in low-calcium fly ashes
Highly active pozzolans		
Condensed silica fume	Consists essentially of pure silica in noncrystalline form	Extremely fine powder consisting of solid spheres of 0·1 μm average diameter (about 20 m^2/g surface area by nitrogen adsorption)
Rice husk ash (Mehta–Pitt process)	Consists essentially of pure silica in noncrystalline form	Particles are generally less than 45 μm but they are highly cellular (about 60 m^2/g surface area by nitrogen adsorption)
Normal pozzolans		
Low-calcium fly ash	Mostly silicate glass containing aluminum, iron, and alkalis. The small quantity of crystalline matter present generally consists of quartz, mullite, sillimanite, hematite, and magnetite	Powders corresponding to 15–30% particles larger than 45 μm (usually 200–300 m^2/kg Blaine). Most particles are solid spheres with average diameter 20 μm. Cenospheres and plerospheres may be present
Natural materials	Besides aluminosilicate glass, natural pozzolans contain quartz, feldspar, and mica	Particles are ground to mostly under 45 μm and have rough texture
Weak pozzolans		
Slowly cooled blast-furnace slag, bottom ash, boiler slag, field burnt rice husk ash	Consists essentially of crystalline silicate materials, and only a small amount of noncrystalline matter	The materials must be pulverized to very fine particle size in order to develop some pozzolanic activity. Ground particles are rough in texture

of hardened concrete through their fine particle size and slow chemical reactions.

The ASTM has separate specifications for air-entraining chemicals (ASTM C 260), and for water-reducing and set-controlling chemicals (ASTM C 494). The latter are divided into seven types: Type A, water-reducing: Type B, set-regarding; Type C, set-accelerating; Type D, water-reducing and set-retarding; Type E, water-reducing and set-accelerating; Type F, high-range water-reducing (superplasticizing); and Type G, high-range water-reducing and set-retarding. Normal water-reducing admixtures should be able to reduce at least 5% of the mixing water for a given consistency of concrete; superplasticizers should be able to reduce the mixing water content by at least 12%.

The mineral admixtures are covered by ASTM C 618, except for ground iron blast-furnace slag which is covered by ASTM C 989. A summary of the different classes of mineral admixtures, their compositions, and physical characteristics is given in Table 3.7.

Since corrosion of reinforcing steel in concrete is a major potential problem with marine structures, the use of *corrosion-inhibiting chemicals,* such as calcium nitrite[5] and sodium thiocyanate,[6] as concrete admixtures has been advocated. Although calcium nitrite has been used commercially, there is a controversy about the long-term effectiveness of corrosion-inhibiting admixtures. For instance, Corbo and Farzam[7] maintain that the concentration of nitrite ions in pore solution decreases with time, and therefore its corrosion-inhibiting effectiveness is likely to be reduced in applications where the ingress of chloride from external sources is unrestricted, as in the case of bridge decks, parking garages, and marine structures. Accordingly, the authors believe[8] that the use of good-quality concrete, such as superplasticized concrete containing silica fume, can prevent steel corrosion more effectively than the use of a corrosion-inhibitor.

MICROSTRUCTURE OF CONCRETE

At the macrostructural level, hardened concrete is a two-phase material composed of hydrated cement paste and aggregate. Since the coarse aggregate particles exercise an important influence on the drying shrinkage, creep, and permeability characteristics of concrete, sometimes it is necessary to think of concrete as a dispersion of coarse aggregate particles in a mortar mixture. Also, the microstructure and properties of the hydrated cement paste in the immediate vicinity of coarse aggregate particles happen to be very different from those of the cement paste in the bulk mortar; therefore it is important to understand these microstructural differences which are now known to have a profound effect on the properties of concrete. A small area surrounding a coarse aggregate is called the *transition zone*; the microstructure of the hydrated cement paste in the transition zone is described below.

The Transition Zone

Although relatively small in area (typically 10 to 50 μm wide around the particles of coarse aggregate), the transition zone between the cement paste and coarse aggregate exercises a far greater influence on the behavior of concrete than is generally appreciated. The reasons why the transition zone is formed in concrete, and why the microstructure and properties of the hydrated cement paste in the transition zone are different from those of the bulk cement paste, are briefly discussed next.

The water content to achieve the desired consistency (e.g., 100–150 mm slump) in an ordinary concrete mixture, without any admixtures present, is generally high enough to cause 'bleeding' when a freshly placed concrete mixture is tamped or vibrated for the purpose of consolidation. Bleeding is a form of segregation. Since water is the lightest component in a fresh concrete mixture it tends to move upwards, whereas the heavier solids tend to settle down. Only a part of the bleed-water reaches the concrete surface where it may be lost by evaporation. Owing to the wall effect, a large amount of bleed-water gets blocked at the aggregate surfaces, especially at the underside of coarse aggregate particles (and reinforcing steel when present). A diagrammatic representation of this phenomenon is shown in Fig. 3.5(a). Failures at the cement paste–aggregate interface result from the weak transition zone (Fig. 3.5(b)).

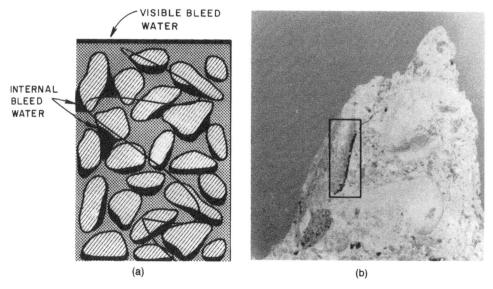

Fig. 3.5. (a) Diagrammatic representation of bleeding in freshly deposited concrete; (b) shear-bond failure in a concrete specimen tested in uniaxial compression (reproduced, with permission, from Ref. 1).

As a result of the much higher than prescribed water/cement ratio, the hydrated cement paste in the transition zone is very porous and weak when compared to the bulk cement paste. It should be noted that, from the hydration reactions involving C_3S and C_2S, a fully hydrated portland cement paste contains approximately 20% to 25% calcium hydroxide by volume of total solids. In addition to high porosity, the large size of the preferentially oriented crystals of calcium hydroxide (Fig. 3.6), which is intrinsically weaker than the other hydration products of portland cement, also contributes to the poor strength of the cement paste in the transition zone.

In ordinary concrete, the transition zone at early ages is usually so weak that it microcracks easily even under the influence of stresses induced by environmental temperature and humidity changes. It is the presence of interconnections between capillary voids in the bulk cement paste and the microcracks in the transition zone which accounts for the general phenomenon that, with a given water/cement ratio, the permeability coefficient of concrete is two orders of magnitude higher than that of the

Fig. 3.6. Scanning electron micrograph of calcium hydroxide crystals in the transition zone (reproduced, with permission, from Ref. 1).

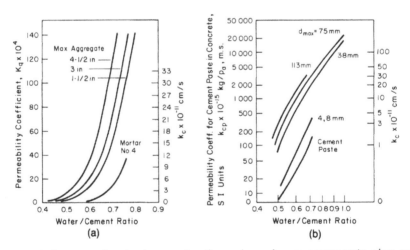

Fig. 3.7. Influence of water/cement ratio and maximum aggregate size on concrete permeability: (a) K_q is a relative measure of the flow of water through concrete in cubic feet per year per square foot of area for a unit hydraulic gradient. ((a), reproduced, with permission, from *Concrete Manual*, 8th edn, US Bureau of Reclamation, 1975, p. 37; (b) adapted, with permission, from Beton-Bogen, Aalborg Cement Co., Aalborg, Denmark, 1979.)

corresponding cement paste. The microcracks in the transition zone of concrete also explain why there is an exponential rise in the coefficient of permeability with increasing water/cement ratio or increasing aggregate size (Fig. 3.7).

In conclusion, the microstructure of concrete is highly complicated since each of the three phases is heterogeneous and complex. Not only is each particle of natural aggregate generally multimineralic and microporous, but also the cement paste in the bulk and in the transition zone contains a heterogeneous distribution of different types and amounts of solids and pores. A further complication in predicting and controlling the microstructure of concrete lies in the fact that the microstructure is subject to changes with time, temperature, humidity, and other environmental effects.

PROPERTIES OF CONCRETE

In modern sea structures the desired properties of concrete are often demanding, and occasionally conflicting. To arrive at optimal solutions, a knowledge of the important properties of concrete is essential. Only some properties of hardened concrete, such as strength, dimensional stability, permeability, and durability are described here, and then only briefly. For a detailed description, standard textbooks on concrete may be consulted.[1,2,9,10]

Strength

Uniaxial compressive strength has traditionally been the most desired property by which the general quality of concrete is evaluated. While the compressive strength of ordinary concrete is assumed to serve as a measure of the other strength types, such as tensile, shear, and flexural, as well as of the elastic modulus and resistance to corrosive and abrasive environments, it is not necessarily an accurate measure of these and other properties of concrete.

Tensile strength determines the onset of cracking, flexural and shear strengths, and fatigue endurance. Owing to the presence of numerous microcracks and voids in ordinary concrete (20–30 MPa 28-d compressive strength), the tensile, flexural, and shear strengths are not high; typically they are of the order of 9%, 15%, and 20% of the uniaxial compressive strength, respectively. Since 2–3 MPa tensile strength is too low to withstand even the tensile stresses induced by drying shrinkage and thermal shrinkage, unreinforced concrete elements are designed to withstand only compressive loads.

According to Gerwick,[11] reinforced and especially prestressed concretes show excellent fatigue in air as long as the concrete is neither cycled into the tensile range to a greater level than half its static tensile strength nor into its compressive range to more than half its compressive strength. These limits are normally readily met in design practice. When submerged, conventional concrete shows a reduction of fatigue endurance, apparently due to high pore pressures generated within the microcracks. Muguruma[12] has reported that dense, low-water/cement-ratio, concretes are not as sensitive to reduction of the fatigue strength under submerged conditions.

Elastic Modulus, Creep, and Shrinkage

The static elastic modulus of ordinary concrete in compression varies from 20 to 30 GPa (3 to 4×10^6 psi). Although before failing in compression concrete exhibits some inelastic (irreversible or plastic) strain, typically of the order of 2000×10^{-6}, it is not considered a ductile material and must be reinforced with steel when designing for tensile, flexural, and impact loads. It should be noted that the inelastic strain is caused by the propagation of microcracks, and is not due to any true plasticity of the material.

A concrete element will also exhibit plastic strain when held for a long period under a stress level that is considerably lower than the failure stress, for instance 50% of the ultimate strength of the material. This phenomenon of gradual increase in strain with time under a sustained load is called creep. Both the drying shrinkage and creep strains in ordinary concrete can range from 500 to

1000×10^{-6}. When the creep strain in concrete is restrained, it manifests itself as a progressive decrease of stress with time. The stress relief associated with the creep strain is generally responsible for the prevention or delay of cracking in concrete elements due to tensile stresses induced by the drying shrinkage. On the other hand, in prestressed elements the creep phenomenon will be responsible for the partial loss of prestress.

In addition to the drying shrinkage, it should be noted that shrinkage strains can also be caused by cooling hot concrete. Ordinary portland cement produces approximately 250–300 kJ/kg heat of hydration within 72 hours of the addition of water. Depending on the cement content of the concrete and the size of concrete element, an adiabatic temperature rise of the order of 30–50°C is therefore not uncommon in massive structures. When cooled to ambient temperature, such structures may experience thermal strains of 200 to 400×10^{-6}, depending on the coefficient of thermal expansion of the concrete, which ranges from a low of about 6×10^{-6} per degree Celsius for limestone and gabbro aggregates, to a high of 12×10^{-6} per degree Celsius for some sandstones, natural gravels, and quartzitic aggregates. High shrinkage strains are critical to concrete because, when restrained, they manifest themselves as high tensile stresses, which frequently lead to concrete cracking. Of course, cracking is also determined by the rate at which a hot concrete is cooled to the ambient condition.

Permeability

The permeability of concrete is one of the properties which is hard to measure and specify, although it plays an important role in determining the durability of materials in corrosive environments, such as seawater. Typical values of coefficients of permeability for ordinary structural concrete (300–350 kg/m^3 concrete, 0·5–0·6 water/cement ratio, 25–37 mm aggregate) are $1–2 \times 10^{-12}$ m/s. The permeability of concrete depends not only on water/cement ratio and aggregate size, but also on consolidation, curing, and good concreting practice. For instance, exposure of hot concrete elements to cool air or cool water immediately after removal of formwork will cause microcracking and will considerably enhance the permeability of concrete. Microcracking in the splashing zone of a concrete structure may be caused by normal cycles of heating and cooling in the climates in which the temperature differences are high between air and seawater, or between night and day.

REFERENCES

1. Mehta, P. K., *Concrete: Structure, Properties, and Materials,* Prentice Hall, Englewood Cliffs, NJ, 1986.
2. Lea, F. M., *The Chemistry of Cement and Concrete,* 3rd edn, Chemical Publishing Co., New York 1971, p. 627.
3. *Handbook of Concrete Admixtures,* ed. V. S. Ramachandran, Noyes Publication, Park Ridge, NJ, 1984.
4. *Concrete Admixtures: Use and Applications,* ed. M. R. Rixom, The Construction Press, New York, 1978.
5. Berke, N. & Rosenberg, A. M., *Transportation Research Record,* No. 1212, Transportation Research Board, 1989.
6. Nmai, C. K. & Corbo, J. M., Sodium thiocyanate and the corrosion potential of steel in concrete, *Concrete International,* **11**(11) (1989), 59–67.
7. Carbo, J. & Farzam, H., Influence of three commonly used inorganic compounds on pore solution chemistry and their possible implications to the corrosion of steel in concrete, *ACI Materials J.,* **86**(5) (1989), 498–502.
8. Carbo, J. & Farzam, H., Letters to the Editor, *Concrete International,* **12**(5) (1990), 13.
9. Mindess, S. & Young, J. F., *Concrete,* Prentice Hall, Englewood Cliffs, NJ, 1981.
10. Neville, A. M., *Properties of Concrete,* Pitman, London, 1981.
11. Gerwick, B. C., *Construction of Offshore Structures,* Wiley–Interscience, 1986, pp. 64–5.
12. Muguruma, H., Low-cycle fatigue behavior of concrete, *Proceedings Gerwick Symposium, on Durability of Concrete in Marine Environment,* ed. P. K. Mehta, Dept. of Civil Engineering, University of California at Berkeley, 1989, pp. 187–212.

Chapter 4
History of Concrete Exposed to Seawater

There is a general perception that concrete is a highly durable material. However, the history of concrete in marine environments shows that no construction material is infallibly durable to seawater, which presents one of the most corrosive natural environments in the world. Many inland structures in Europe built by the Romans with lime-pozzolan cement have withstood the ravages of time for two thousand years. On the other hand, few such ancient structures have survived on the sea coasts. We will never know about those numerous old structures that have fully succumbed to the destructive forces of the ocean.

In the concrete literature, there are two 150–200-year-old reports on the durability of concrete to seawater.[1,2] However, since the 1900s several international organizations have held conferences on the subject and published proceedings, for instance, the International Association for Testing Materials (1909, 1912, 1927), and the Permanent International Association of Navigation Congresses (1923, 1926, 1931, 1949). In America, the results of a comprehensive survey on the performance of concrete sea structures in the United States were reported by Wig and Ferguson,[3] and a state-of-the-art report was published by Atwood and Johnson.[4] More recently, the American Concrete Institute has published two ACI/CANMET conference proceedings.[5] Also, there are two noteworthy publications based on the Scandinavian experience.[6,7] As described next, a review of the published literature on the history of the performance of concrete in seawater also provides a fascinating story of the history of the development of cement and concrete technology.

In 1756, John Smeaton,[1] a British engineer who reportedly was the first person to call himself a *civil engineer,* was commissioned to build a lighthouse on the Eddystone Rock situated at the western outlet of the English Channel. Owing to the severity of wave action and chemical attack by seawater, he wanted a stronger and more durable cement than the traditional blends of slaked lime and pozzolans which were being used at that time in Europe since the Roman period. The traditional cement for making a water-resisting mortar consisted of two parts of slaked lime mixed with one part of a zeolitic pozzolan (West German trass from Andernach), beaten together to form a paste with water. Since Smeaton did not find this cement to have a satisfactory resistance to seawater, he experimented with several samples of lime from different origins. He found that the lime which produced the best mortar came from calcination of a limestone containing a considerable amount of clay matter. According to Lea,[8] this was probably the first time that the properties of hydraulic lime, a precursor of portland cement, were recognized. The Eddystone Lighthouse (Fig. 2.2) is thus an important landmark in the history of the development of portland cement as well as of marine transportation. The lighthouse lasted for approximately 120 years until the failure of its foundation.

In 1818, L. J. Vicat's[2] investigations in France on hydraulic lime led to the production of synthetic hydraulic lime by calcination of an artificial blend of a high-purity limestone and clay. This technique was the forerunner of the technology of modern portland cement, which was patented in 1824 by Joseph Aspdin, a British builder. Subsequently, in harbor and coastal structures the traditional construction materials, such as timber and natural stone, were gradually replaced with portland cement concrete, which was found to be more versatile and easier to work with. The advent of reinforced concrete at the close of the nineteenth century provided the real impetus for the rapid expansion of marine structures all over the world, and this contributed further to the growth of the cement and concrete industries. European nations in the nineteenth century, especially Britain and France, played a leading role in the development of cement and concrete technology as well as marine technology, because ocean navigation was essential to the spread of the industrial revolution at home and colonization abroad.

According to Lea,[8] Vicat was the first to propose that the chemical attack of seawater on lime mortar was mainly a result of interaction between the uncombined $Ca(OH)_2$ in the mortar and magnesium sulfate present in seawater. Vicat began his meticulous experimental studies in 1812 when, according to him, there was a 'chaos of opinion' on the subject. The first results of his work were published in 1818, but the complete work entitled, 'Research on the Causes of Physical Destruction of Hydraulic Mortars by Seawater' (in French), was published in 1857. For his work Vicat received two prizes of 2000 francs each, which had been offered by the Société d'Encouragement pour l'Industrie Nationale (Society for Encouragement of National Industry) for the best work on durability of cement mortars for marine construction. Vicat made this profound observation:

> On being submitted to examination, the deteriorated parts (of the mortar) exhibit much less lime than the others; what is deficient then, has been dissolved and carried off; it was in excess in the composite cement mixture. Nature, we see, labors to arrive at exact proportions, and to attain them corrects the errors of the hand which had formulated the original proportions. Thus the effects which we have just described, and in the case alluded to, become more marked the further we deviate from these exact proportions.

In addition to the deterioration of concrete in seawater, some failures were experienced in southern France as well as in southern Algeria due to decomposition of mortars and concretes in sulfate soils. Replacement of lime–pozzolan mixtures with portland cement did not help since the hydration of portland cement liberated considerable free $Ca(OH)_2$. Another French engineer, J. Bied, following Vicat's ideas, invented a non–calcium silicate cement which, on hydration, does not produce any $Ca(OH)_2$. Popularly known as *high-alumina cement* or *ciment fondu*, the principal compound in this cement is monocalcium aluminate. The cement proved excellent for its resistance to seawater and other sulfate-bearing waters. However, it suffers from loss of strength on exposure to warm and humid conditions, and therefore its use for structural work is now banned in most countries of the world.

In 1880, an investigation by Prazier into the causes of concrete deterioration at Aberdeen Harbor, Scotland provided a confirmation of Vicat's findings. Prazier concluded that hydraulic cements used in concrete exposed to seawater lost lime and absorbed magnesium from seawater by an ion-exchange reaction which caused the degradation of concrete. In 1924, based on a comprehensive review of the previous 100 years of worldwide experience with concrete in seawater, Atwood and Johnson[4] essentially reconfirmed what Vicat had discovered more than one hundred years before. The oldest cementing material cited by the authors—an AD 60 lime–pozzolan mortar—when compared to five recent mortars (1886–89) showed that after the action of seawater (i.e. after removal of a part of the original lime present) the residual lime content in all six specimens was similar.

More recent investigations of deteriorated marine structures have shown that in permeable concrete, in addition to the magnesium sulfate attack, the carbonic acid attack (from dissolved CO_2 in seawater) is also an important factor. According to Feld,[9] in 1955, after 21 years of service the concrete piles and caps of the trestle bends of the James River Bridge in Newport News, Virginia, required a $1·4 million repair involving 70% of the 2500 piles. Similarly, after 25 years of service, in 1957, approximately 750 precast piles near Ocean City, New Jersey, had to be repaired because of severe loss of mass; some of the piles had been reduced from their original 550 mm diameter to 300 mm. In both cases the disintegration and loss of material was primarily attributed to abnormally high concentrations of dissolved CO_2 in seawater; the pH of the seawater was found to be close to 7 instead of 8·2 to 8·4, which is the pH value of normal seawater.

Feld[9] described another case of concrete deterioration, highlighting the importance of permeability. Two of the ten piers of the Shrewberry River Bridge, New Jersey, built in 1913, showed signs of failure after about one year of service. The removal of the deteriorated concrete, which had been placed by tremie, showed alternate layers of good concrete and mushy concrete (putty-like material containing crystalline salts). The accumulation of laitance layers, 50–75 mm thick, was common in 5 m depths of placement.

It seems that the failure to remove completely the laitance layers resulted in permeable seams within the concrete, which were easily exposed to seawater attack.

Mehta and Haynes[10] described a field investigation involving 18 unreinforced concrete test blocks ($1 \cdot 8 \times 1 \cdot 8 \times 1 \cdot 1$ m) which, in 1905, were partially submerged in seawater at San Pedro harbor near Los Angeles, California. Examined in 1972, after 67 years of exposure to seawater, it was found that the dense concrete blocks (1:2:4 parts cement, sand, and gravel, respectively) were in excellent condition although some of them had been made with a high-C_3A portland cement (14% C_3A). On the other hand, lean concrete blocks (1:3:6) had lost some mass and had a soft surface, which was covered with marine growth (Fig. 4.1). Mineralogical analyses of the deteriorated concrete showed significant amounts of magnesium hydroxide, gypsum (calcium sulfate), ettringite, aragonite (calcium carbonate), and hydrocaluminate (carboaluminate hydrate). The original products of portland cement hydration, namely calcium silicate hydrate and calcium hydroxide, had disappeared as a result of magnesium sulfate and CO_2 attack. From long-time field tests Regourd[11] and Gjørv[12] reported similar

Fig. 4.1. Marine growth on lean concrete block exposed to seawater for 67 years (reproduced, with permission, from Ref. 10).

results. From the selected case histories described above, it was concluded that permeable concrete would eventually deteriorate as a result of chemical attack by seawater and that the permeability of concrete is more important in determining the long-term durability than differences in the composition of the portland cement.

Field experience with durability of marine structures shows that chemical attack is not the only problem. Concrete in breakwaters, jetties, groynes, piles, and other marine structures must resist enormous wave pressures and impacts, especially during storms. Idorn[13] (see Table 2.3) cites a number of breakwater failure cases in the Mediterranean during the 1970s: Antalya in Turkey (1971), Arzew el Djedid in Algeria (1979), Tripoli in Libya (1977, 1980), Sines in Portugal (1978), Bilbao and San Ciprian in Spain (1980). Insufficient tensile and flexural strengths of unreinforced mass concrete to resist the wave action contributed to these failures. Consequently, most structures today are reinforced with steel. However, this adds another complication to the durability problem because steel is subject to corrosion in salt water. Furthermore, many structures are located in cold climates where they are exposed to the action of freezing and thawing cycles. A few case histories on the performance of reinforced structures, both in mild and cold climates, are described next.

In 1912, in the Los Angeles harbor, precast reinforced concrete piles were driven to support a 800-m wharf.[14] After 12 years of service some piles showed longitudinal cracks at the mid-tide level, leading eventually to spalling of concrete. The concrete below the low-tide level remained sound. The authors speculated that microcracks from sulfate attack could have been responsible for the permeation of seawater, which subsequently caused corrosion of the reinforcement. On the basis of experience elsewhere it is possible that other sources of microcracking of concrete such as thermal gradients or pile driving against high resistance, may have been overlooked. Fluss and Gorman[15] surveyed the condition of 46-year-old reinforced concrete piles and girders of the San Francisco ferry building, constructed in 1912. It is reported that the concrete contained a high-C_3A portland cement (14–17% C_3A), and a high cement content (approximately $400\,kg/m^3$). Most of the piles were found in good condition. Some of the piles and transverse girders of two piers were cracked. The authors were of the opinion that poor workmanship and excessive deflection under load might have been the sources of microcracking in the concrete, which exposed the reinforcing steel to the corrosive action of seawater.

The results of a 1953–55 survey of more than four hundred, 20–50-year-old, coastal structures in Denmark were published by Idorn[6] in 1967. About 40% of the surveyed structures showed some deterioration. Among the severely deteriorated ones were Pier No. 7 of the Oddesund Bridge, a highway bridge in North Jutland, and a concrete barrier at the Lim Fjord. The concrete from Pier No. 7 of the Oddesund Bridge showed cracking and strength loss due to a combination of freezing and thawing cycles, alkali–silica reaction, and sulfate attack. The historical record of the structure indicated initial cracking of the caisson due to thermal stresses and this, according to Idorn, increased the permeability of concrete to seawater and set the stage for subsequent chemical attack. The pier had to be repaired after 8 years of service. The second structure, a highway bridge in North Jutland, showed severe cracking and spalling of concrete in the bridge piers. The loss of mass was maximum at the mid-tide level, giving a typical hour-glass shape to the columns. Corrosion of the reinforcement was rampant in longitudinal girders. Analysis of concrete samples showed evidence of poor quality (high water/cement ratio). Also, other deleterious phenomena responsible for concrete cracking, such as freezing–thawing and alkali–silica reaction, were superimposed on the reinforcement corrosion. The third structure, Groin 72 of the Lim Fjord, was made of highly permeable and low-strength concrete ($220 \, kg/m^3$ cement content), which was unsuitable to withstand repeated cycles of wetting–drying, in addition to strong wave action (impact of sand and gravel in the surf). Some of the concrete blocks had totally disintegrated in the course of 20 years of exposure to seawater.

In 1962–64, along the Norwegian seaboard over seven hundred concrete structures were surveyed by Gjørv.[7] Of these structures, 60% were reinforced concrete wharves of the slender-pillar type made with tremie-cast concrete. At the time of the survey about two-thirds of the structures were 20–50 years old. The author reported that, in general, concrete pillars were in good condition below the low-tide level and above the high-tide level. However, in the splashing zone many pillars showed signs of damage; approximately 14% had their area of cross-section reduced by 30% or more, and 24% had a 10–30% reduction. The deterioration of concrete in the splashing zone was ascribed mainly to the effect of freezing and thawing cycles on non-air-entrained concrete. Evidence of heavy damage due to corrosion of reinforcing steel was found on 20% of the deck beams.

Mehta and Gerwick[16] reported that in 1980 several spandrel beams of the San Mateo–Hayward bridge near San Francisco, California, were found to be in need of repair at a great expense because of serious cracking of concrete, presumably from corrosion of the embedded steel (Fig. 4.2). Seventeen years earlier a high-quality concrete mixture ($370 \, kg/m^3$ cement content, 0.45 water/cement ratio) had been used for making the beams. Some of the beams were precast and steam-cured while the others were cast-in-place and naturally cured. No corrosion and cracking were discovered in the cast-in-place beams; however, all the steam-cured beams showed cracking and corrosion at the underside and windward faces, which were directly exposed to

Fig. 4.2. Concrete damaged by corrosion of the reinforcing steel in precast spandrel beams of San Mateo–Hayward Bridge (reproduced, with permission, from Ref. 16).

seawater spray. The authors proposed that because of the massive size of the beams (approximately $8 \times 3.75 \times 1.8$ m) a combination of heavy reinforcement and differential cooling rates during the manufacturing process (subsequent to the heating cycle in steam curing) could have produced invisible microcracks. Later, the microcracks must have become continuous in that part of the structure which was exposed to severe weathering action. It seems that under the environmental conditions presented typically by the marine environment, a concrete which originally may have been impermeable can subsequently become permeable, and therefore vulnerable to a corrosion–cracking–corrosion cycle leading eventually to serious structural damage.

Recently, Gjørv and Kashino[17] were able to conduct a detailed investigation on durability of concrete from a 60-year-old pier before the structure was demolished. A non-air-entrained concrete mixture, with $350 \, kg/m^3$ cement content and 0.53 water/cement ratio (30 MPa design strength) had been prescribed for use. Concrete cores from the deck showed no evidence of sulfate or chemical attack, and excellent strength and impermeability after 60 years of exposure to seawater (42–45 MPa compressive strength, $2.5–8 \times 10^{-13} \, kg/Pa$ ms coefficient of permeability). Since a portland cement with high C_3A content had been used for making this concrete, the data provide a reconfirmation that permeability rather than cement composition is the most important parameter for the chemical resistance of concrete to seawater. For instance, the sulfate attack associated with the formation of ettringite was confined to the surface of the concrete up to a depth of 5 mm. The authors observed some microcracks, 10–35 mm deep, but these were attributed to thermal stresses during construction. The depth of carbonation was low (1–7 mm) in the submerged zone and the tidal zone, but was relatively high (up to 24 mm) in the upper part of the tremie-cast concrete pillars. This is consistent with the view of many researchers that the rate of diffusion of gases into saturated concrete drops off sharply.

Gjørv and Kashino[17] reported that in the lower part of the deck chloride ions had penetrated to a depth of more than 80 mm, and the chloride content at the steel bars was $0.05–0.1\%$ by weight of concrete. In the fully submerged part of the structure, the corresponding chloride content was $0.3–0.35\%$. The pier had been subjected to frequent repairs during its service life, mainly an account of cracking from frost action and corrosion of reinforcing steel; the first repair was carried out after only 10 years' exposure to seawater and the next one after 16 years. The authors observed that cleaning and recasting of recracked concrete in spalled areas had not stopped subsequent corrosion because, after a few years, new cracks appeared in the adjacent areas. Although deck beams at the inner and most protected part of the pier were generally in good condition, at the outer part, which was more exposed to the weathering action of seawater, all deck beams showed visible deterioration of concrete due to rebar corrosion. Several pillars were also found to have cracked due to rebar corrosion in the upper part of the structure (above the tidal zone); however, there was no evidence of cracking and rust staining in the fully submerged parts.

In conclusion, although reinforced concrete with low permeability shows adequate chemical resistance to seawater, in cold climates an unprotected concrete (non-air-entrained) is likely to crack as a result of frost action and will subsequently deteriorate further due to corrosion of reinforcing steel. This conclusion was recently confirmed by Khanna et al.,[18] who reported the results of an investigation on deteriorated reinforced concrete piles from the Rodney Terminal at Saint John in New Brunswick, Canada. The investigation revealed that thermal cracks during the pile manufacture were precursors to cracking due to freezing and thawing cycles (since there was lack of adequate air entrainment in concrete). Thus corrosion-cracking was judged to be the secondary cause for deterioration of concrete.

Many investigators with long-term professional experience on durability of concrete structures in marine environments have observed that in hot climates the electrochemical corrosion of reinforcing steel in concrete is greatly enhanced. As a result, a concrete mixture that performs adequately in a cold climate may not perform satisfactorily in hot climates. According to Gerwick[19] the Dubai tunnel, built in 1973–75, had to be completely repaired in 1986 at a cost twice that of the original construction cost. The Middle East environment is so severe (hot days followed by cold nights) that some consider the area to be a full-scale accelerated testing laboratory. In the Seawater Cooling Canal Project, repairs to concrete as a result of corrosion-cracking had to be undertaken even before the canals were in full service. The new dry dock at Abu Dhabi is reportedly also severely damaged; the alternate wetting and drying in the hot climate seems to constitute an especially aggressive condition.

As a result of inspection of coastal marine structures in the Gulf area during the period 1974 to 1986, Normand[20] found that many reinforced concrete structures had suffered damage, ranging from minor surface weathering to major spalling, due to the reinforcement corrosion. The damage had occurred within a relatively short service life of 3 to 10 years following the construction. Compared to today's standards, these structures were built with a relatively poor-quality concrete (<30 MPa) and 50 mm specified cover thickness. In some cases, particularly where local pitting was evident, the rebars had suffered a significant loss of their cross-sectional area. In the case of unreinforced mass concrete structures, Normand[20] discovered that the overall deterioration with similar quality concrete was far less, although some salt weathering at the surface was evident. Consequently, the author recommended that with marine structures the possibility of using mass concrete should always be examined before the use of reinforced concrete is specified. With a much better knowledge of the causes for lack of durability, which has been incorporated into various codes of recommended practice, the author felt that the reinforced concrete structures built during the 1980s should have a satisfactory performance record.

In fact, Normand's conclusion is confirmed by the experience with offshore concrete platforms in the North Sea, which are constructed with high-quality (40–60 MPa) concrete. These structures are also exposed to temperature extremes, although, instead of the hot-dry or hot-wet conditions in the Gulf area, cold weather conditions predominate in the North Sea.

The oldest offshore concrete platform in the North Sea, Ekofisk, has been in service for more than 20 years and only 7 of the 20 concrete platforms have been in service for more than 10 years. Hoff[21] reviewed the performance of the North Sea concrete platforms with respect to freezing and thawing damage, corrosion of reinforcement, chemical and marine organism attacks, and impact loading. Although the service life so far has been relatively short, the author concluded that the overall performance of concrete in the existing North Sea structures has been exceptionally good, and the performance is likely to remain satisfactory for many years to come.

REFERENCES

1. Smeaton, J., *A Narrative of the Building of the Eddystone Lighthouse*, 1793; also, Lea, F. M., *Chemistry of Cement and Concrete*, Chemical Publishing Co., New York, 1971.
2. Vicat, L. J., *Treatise on Calcareous Mortars and Cements*, French version of 1818, translated by J. T. Smith in 1837.
3. Wig, R. J. & Ferguson, L. R., What is the trouble with concrete in seawater?, *Eng. News Record*, **79** (1917).
4. Atwood, W. G. & Johnson, A. A., The disintegration of cement in seawater, *Trans. ASCE*, **87** (1924), 204–30.
5. *Performance of Concrete in Marine Environment*, ed. V. M. Malhotra, ACI-SP 65 (1980), and ACI-SP 109 (1988), 627 pages.
6. Idorn, G. M., *Durability of Concrete Structures in Denmark*, Technical University, Denmark, 1967, 208 pages.
7. Gjørv, O. E., *Durability of Reinforced Concrete Wharves in Norwegian Harbors*, Ingenior fo-plaget A/S, Oslo, 1968, 208 pages.
8. Lea, F. M., *The Chemistry of Cement and Concrete*, 3rd edn, Chemical Publishing Co., New York 1971, pp. 5–10.
9. Feld, J., *Construction Failures*, John Wiley, New York 1968, pp. 251–5.
10. Mehta, P. K. & Haynes, H., Durability of concrete in seawater, *J. ASCE Structural Division*, **101**(ST8) (1975), 1679–86.
11. Regourd, M., The action of seawater on cements, *Annales de l'Institut Technique du Batiment et des Travaux Publics*, No. 329 (1975), 86–102.
12. Gjørv, O. E., Long-time durability of concrete in seawater, *ACI J. Proc.*, **68** (1971), 60–7.
13. Idorn, G. M., Marine concrete technology—viewed with Danish eyes, *Proceedings Gerwick Symposium on Durability of Concrete in Marine Environment*, ed. P. K. Mehta, Dept of Civil Engineering, University of California at Berkeley, 1989, p. 40.

14. Wakeman, C. M., Dockweiler, E. W., Stover, H. E. & Whiteneck, L. L., Use of concrete in marine environment, *ACI J. Proc.*, **54** (1958), 841–56.

15. Fluss, P. J. & Gorman, S. S., Discussion of the paper by Wakeman *et al.* (Ref. 14), *ACI J. Proc.*, **54** (1958), 1309–46.

16. Mehta, P. K. & Gerwick, B. C., Cracking-corrosion interaction in concrete exposed to marine environments, *Concrete International*, **4**(10) (1982), 45–51.

17. Gjørv, O. E. & Kashino, N., Durability of a 60-year old reinforced concrete pier in Oslo Harbor, *Materials Performance*, **25**(2) (1986), 18–26.

18. Khanna, J., Seabrook, P., Gerwick, B. C. & Bickley, J., Investigation of distress in precast concrete piles at Rodney Terminal, *Performance of Concrete in Marine Environment*, ed. V. M. Malhotra, ACI SP-109, 1988, pp. 277–320.

19. Gerwick, B. C., Pressing needs and future opportunities in durability of concrete in the marine environment, *Proceedings Gerwick Symposium on Durability of Concrete in Marine Environment*, ed. P. K. Mehta, Dept. of Civil Engineering, University of California at Berkeley, 1989, pp. 1–5.

20. Normand, R., Review of the performance of concrete coastal structures in the Gulf area, *Proceedings, International Conference on Concrete in the Marine Environment*, The Concrete Society, London, 1986, pp. 101–13.

21. Hoff, G. C., The service record of concrete offshore structures in the North Sea, *Proceedings, International Conference on Concrete in the Marine Environment*, The Concrete Society, London, 1986, pp. 131–42.

Chapter 5
Causes of Deterioration of Concrete in Seawater

The physical and chemical causes of concrete deterioration are classified in Fig. 5.1 and Fig. 5.2, respectively.[1] From the case histories of the performance of concrete in seawater, as described in the previous chapter, it is clear that many of the physical and chemical causes of concrete deterioration are present in the marine environment. This is diagrammatically illustrated by Fig. 5.3, from which it can be concluded that concrete structures in the ocean environment are exposed to one of the most hostile natural environments in the world.

It should be noted that classification of the causes of concrete deterioration into neat categories has a limited value. Failure analysis of a concrete which is found to be in an advanced state of degradation, generally shows a maze of interwoven chemical and physical causes of deterioration at work. This is because many physical and chemical phenomena are usually interdependent and mutually reinforcing. For instance, expansion and microcracking due to the physical effect of pressure from salt crystallization in a permeable concrete will increase the permeability further and pave the way for deleterious chemical interactions between seawater and cement hydration products. Similarly, chemical decomposition and leaching of the constituents of hydrated cement paste give rise to detrimental physical effects, such as an increase in porosity and therefore a loss in strength.

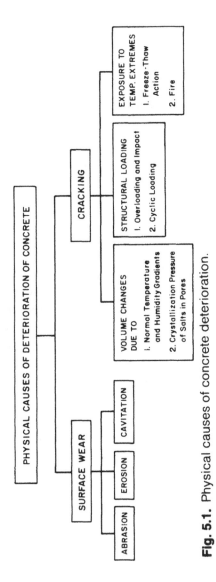

Fig. 5.1. Physical causes of concrete deterioration.

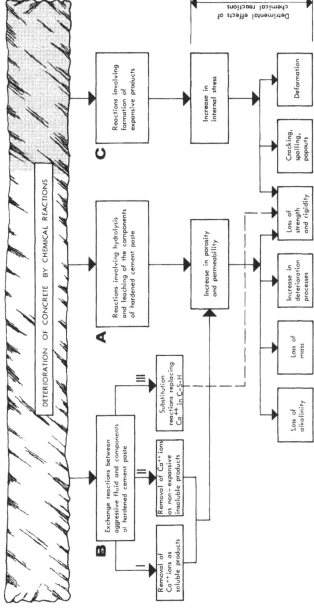

Fig. 5.2. Types of chemical reactions responsible for concrete deterioration. A, Soft–water attack on calcium hydroxide and C-S-H present in hydrated portland cements: B(I), acidic solution forming soluble calcium compounds such as calcium chloride, calcium sulfate, calcium acetate, or calcium bicarbonate; B(II), solutions of oxalic acid and its salts, forming calcium oxalate; B(III), long-term seawater attack weakening the C-S-H by substitution of Mg^{2+} for Ca^{2+}; C, (1) sulfate attack forming ettringite and gypsum, (2) alkali–aggregate attack, (3) corrosion of steel in concrete, (4) hydration of crystalline MgO and CaO.

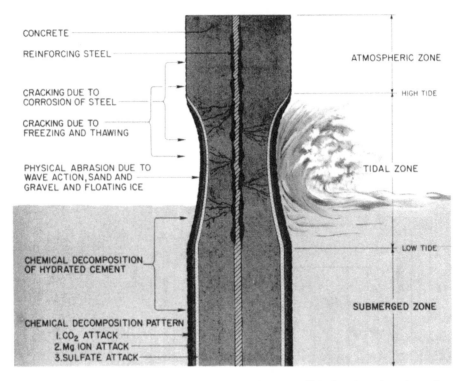

Fig. 5.3. Physical and chemical processes responsible for deterioration of a reinforced concrete element exposed to seawater.

However, categorization of the processes of concrete deterioration into physical and chemical, as shown in Figs 5.1 and 5.2, does serve a purpose. It makes it easier to look at the various phenomena involved, one at a time, for the purpose of understanding the causes and, consequently, the control of the causes. However, for field practice it would be a mistake to ignore the mutual interdependence and synergistic effects, especially when making predictions about the service life of concrete structures in a hostile environment.

It seems that the physical–chemical processes of seawater attack manifest themselves either as cracking of concrete or as loss of mass. Although there are several phenomena capable of causing cracks in concrete, all of them are not equally important. In field practice it is observed that concrete structures deteriorate mostly on account of a combination of stresses generated by cycles of heating and cooling, wetting and drying, freezing and thawing, and corrosion of the reinforcing steel. With reinforced concrete structures, the electrochemical phenomenon of corrosion of the embedded steel is almost always associated with cracking and spalling of the concrete cover. Structures in cold climates may be subjected to numerous cycles of freezing and thawing every winter, and concrete containing inadequate air-entrainment would normally undergo expansion and cracking. In hot and arid regions, concrete in the splashing zone may crack owing to cycles of wetting and drying as well as heating and cooling.

Many concrete structures built before the 1930s were probably made with relatively permeable concrete mixtures and were, therefore, susceptible to various types of chemical attack in seawater, such as magnesium sulfate attack, CO_2 attack, and alkali-aggregate expansion. It should be noted that the chemical decomposition of hydrated cement paste caused by magnesium sulfate and CO_2 attacks leads to strength loss which, in turn, results in loss of mass from erosion of concrete by wave action. Modern marine structures, such as the concrete platforms in the North Sea, are built with concrete mixtures that are relatively impermeable and therefore less vulnerable to chemical attack when compared to old structures.

The mechanisms underlying some of the harmful phenomena mentioned above will now be briefly discussed in the order of their importance to the durability of concrete in the marine environment. An understanding of the causes of deterioration will provide a logical background for their control, which is the key to current recommended practice for durable concrete in the marine environment (Chapter 6, 7).

CORROSION OF REINFORCING STEEL

The corrosion of steel in concrete is an electrochemical process, the corrosion cells being generally formed because of concentration differences in ions and gases in the vicinity of the metal. Normally, reinforcing steel has a thin, $FeO \cdot OH$ film on the surface, which renders the steel passive to the corrosion process. The protective film is stable in the alkaline environment of hydrated portland cement, which generally has a pH > 13. The passivity of the film is destroyed either by a drop in the alkalinity of the contact environment to a pH level below 11 or in the presence of chloride ions. In permeable concrete, carbonation is often responsible for the lowering of pH. The concrete in modern marine structures is essentially impermeable, and therefore the carbonation of cement paste is seldom a matter for concern. Since seawater contains a high concentration of chloride ions, a common cause for the breakdown of the local passivity of reinforcing steel is, therefore, penetration of chloride ions to the steel surface. Chloride concentrations of the order of 0·6 to 0·9 kg/m³ in concrete, or 300–1200 g/liter in the pore fluid, are reported to be sufficient to cause dissolution of the passive film. When the passivity of steel becomes partly or completely broken, the electrochemical potential locally becomes more negative. In other words, the area becomes anodic with respect to another area of steel which continues to remain passive and, therefore, acts as the cathode.

The chemical changes occurring at the anodic and cathodic areas are shown in Fig. 5.4(a). In addition to the passivity breakdown, therefore, there are two other requirements which must be simultaneously fulfilled in order for the corrosion process to proceed at a significant rate. First is the continuous availability of

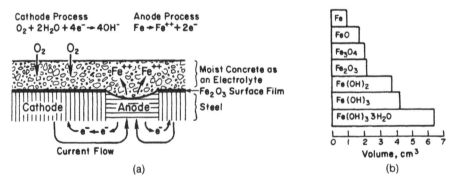

Fig. 5.4. (a) Anodic and cathodic reactions in the corrosion of steel in concrete. (b) Volumetric expansion as a result of oxidation of metallic iron.

oxygen and water at the cathode, and second is the electrical conductivity of concrete (it should be noted that saturated concrete can act as an electrolyte). Significant corrosion rates of steel in concrete are not observed as long as the electrical resistivity of the concrete remains above $50\text{--}70 \times 10^3\ \Omega$ cm.

Depending on the oxidation state of iron, the transformation of metallic iron to rust may be accompanied by a considerable increase in volume—as large as 600%—and this volumetric increase may be the principal cause of concrete expansion and cracking (Fig. 5.4(b)). In a comprehensive report based on an experimental investigation on the large number of intermediate and final products formed during steel corrosion, Figg[2] made the following observations. Akaganeite (beta $FeO \cdot OH$) is invariably associated with the presence of chloride ions (viz. in marine conditions), particularly at intermediate porosities where the relatively slow-moving chloride ions are able to compete effectively against the fast-moving but less available oxygen for anodic Fe^{2+} sites. At lower porosities, the chloride seems to be 'squeezed out' and magnetite (Fe_3O_4) becomes the predominant corrosion product. It may be noted that lepidocrocite (gamma $FeO \cdot OH$) and maghemite (gamma Fe_2O_3), are the usual forms of stable oxidation products on a steel surface in a chloride-free aqueous environment. Chloride environments are highly corrosive to the reinforcing steel because, first, the corrosion process is initiated by transformation of the stable oxidation products present in the protective film to soluble chlorocomplexes of iron and, second, the process is sustained when chloride ions are further attracted to the steel by the positively charged Fe^{2+} at the anode.

The chlorocomplexes, such as $3Fe(OH)_2 \cdot FeCl_2$, appear to be the intermediate products in the formation of akaganeite or other voluminous rust products of gelatinous character. According to Figg,[2] and Mehta,[3] analogously to the internal hydraulic pressure caused by freezing of concrete, or by swelling of the alkali–silicate gel formed as a result of alkali–silica reaction, the hydraulic pressure development, via swelling of gelatinous rust products, should be considered as a possible mechanism of expansion and cracking of concrete from the corrosion of reinforcing steel.

Control of Factors Influencing the Corrosion of Reinforcing Steel

From a consideration of mechanisms underlying the expansion and cracking of concrete due to corrosion of reinforcing steel, as discussed above, it appears that the chloride content of concrete, the availability of oxygen and moisture at the surface of the steel, and the electrical resistivity of the concrete are the important factors controlling the phenomenon.

In field practice, it seems that the penetration of chlorides into hardened concrete occurs more readily by permeation (the permeability of concrete being dependent on water/cement ratio, curing conditions, degree of microcracking, depth of cover, etc.) than by the extremely slow process of diffusion through the saturated cement paste. Based on laboratory or short-term field experiments, the differences in cement types have at times been emphasized as a possible solution to the chloride corrosion problem. For instance, it is suggested that a reinforced concrete element made with a high-C_3A portland cement will perform better in seawater because of the ability of cement to bind chlorides in the form of Friedel's salt, $C_3A \cdot CaCl_2 \cdot 11H_2O$. This suggestion is unsound because, as a result of chemical attack by sulfate ions or CO_2, Friedel's salt will eventually decompose and the chloride will be released again. A lesson can be learnt from the high concentration of chloride and sodium ions in seawater, i.e. these ions are difficult to complex in nature, and cannot be held indefinitely as insoluble compounds in an aqueous environment.

In conclusion, a high-quality concrete mixture (low permeability) and good concreting practice (proper consolidation and curing, adequate depth of concrete cover) appear to be the best safeguards against chloride penetration into concrete in the marine environment. However, if for some reason the chloride concentration at the surface of the reinforcing steel exceeds a certain threshold value (e.g., $0.6 \, kg/m^3$), the passivity of the steel will be destroyed; then the availability of dissolved oxygen at the steel surface and the electrical resistivity of concrete will become the controlling factors for the rate at which corrosion will take place.

According to Gjørv,[4] the availability of oxygen at the surface of the reinforcing steel depends on several factors. First, it should be noted that the concentration of gaseous oxygen in air is approximately 210 ml/liter, whereas the maximum concentration of dissolved oxygen in seawater is only 5–10 ml/liter. For the oxygen to take part in the electrochemical process at the cathode, it must be in a dissolved state; however, the rate of oxygen diffusion in water is extremely poor. Although both the permeability of concrete (which is governed by the water/cement ratio) and the thickness of the concrete cover affect the oxygen availability at the steel surface, Gjørv[4] feels that the degree of water saturation is a more important factor. He cites Tuutti's work according to which the oxygen flux in a portland cement concrete (0·67 water/cement ratio) was reduced by a factor of about 5 when the degree of water saturation increased from 50% to 100%.

The effect of water/cement ratio and cover thickness on the oxygen diffusion rate in water-saturated concrete is shown in Fig. 5.5. The data show that with a given thickness of cover, a decrease in water/cement ratio of concrete from 0·6 to 0·4 reduced the oxygen flux by a factor of about 2; and with a given water/cement ratio, an increase in the cover thickness from 10 to 70 mm reduced the oxygen flux by a factor of about 2·5. Also, with a given water/cement ratio and a given thickness of the cover, the oxygen

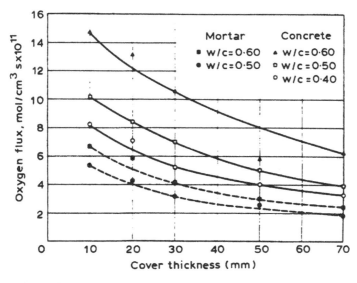

Fig. 5.5. Effect of cover thickness and water/cement ratio on oxygen flux (reproduced, with permission, from Ref. 5).

flux in mortars was less than in the corresponding concretes by a factor of about 2·5. This is attributable to differences in the permeability of the cement paste–steel transition zone between mortars and the corresponding concrete mixtures.

From the standpoint of corrosion of reinforcing steel, on the one hand an increasing degree of water saturation of concrete has a beneficial effect by reducing the rate of oxygen diffusion; on the other hand, however, this can be harmful because the electrical conductivity of the concrete—another requirement which must be fulfilled for the occurrence of the electrochemical phenomenon—increases sharply with the increasing degree of saturation. The effect of degree of water saturation on the electrical resistivity of concrete is shown in Fig. 5.6. The data show that on reducing the degree of water saturation from 100% to 20%, the electrical resistivity of concrete increased from about $7 \times 10^3 \, \Omega$ cm to $6 \times 10^6 \, \Omega$ cm. According to Gjørv,[4] observations from existing

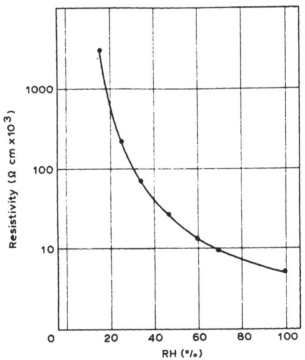

Fig. 5.6. Effect of degree of water saturation on the electrical resistivity (reproduced, with permission, from Ref. 6).

concrete structures indicate that, in practice, corrosion of embedded steel hardly represents a problem as long as the electrical resistivity exceeds a threshold value of $50-70 \times 10^3 \, \Omega$ cm. It can be anticipated that both the permeability of concrete and the ionic concentration in the pore solution will have a significant influence on the electrical resistivity. Accordingly, Vennesland and Gjørv[7] found that the addition of a highly active pozzolanic material, namely 10% condensed silica fume by weight of cement, to a concrete mixture containing $400 \, kg \, cement/m^3$, increased the resistivity by a factor of 5.

In conclusion, even when the passivity of reinforcing steel is destroyed by chlorides, CO_2 or other destructive agents, corrosion of the steel will not occur if the concrete element is in a dry environment (i.e. if the degree of water-saturation is below 80–90%). This is because, for corrosion to take place, the electrical resistivity must be less than a certain threshold value, which cannot be reached in an unsaturated concrete. With increasing degrees of saturation up to 100%, however, the electrical resistivity of concrete will no longer be a limiting factor for corrosion. Now, the corrosion will be controlled by the availability of oxygen at the steel surface. Since the rate of diffusion of dissolved oxygen in fully saturated concrete is extremely slow, it is not surprising that maximum corrosion of a reinforced concrete element occurs where alternate wetting and drying conditions are predominant.

Designers assume that all reinforced structures will crack in service. Consequently, there is considerable controversy in the published literature regarding the relationship between crack width in reinforced concrete cover and corrosion. Most codes of recommended practice specify maximum permissible crack widths, which vary with thickness of the cover over steel reinforcement. This leaves one with the impression that perhaps there is a simple relationship between crack widths and corrosion.

Both theoretical considerations and results from experimental as well as field studies indicate that a simple relationship between crack width and corrosion does not exist. Comprehensive studies by Houston et al. and Schiessel, as cited by Gjørv,[4] support this position. From an extensive study, Beeby[8] came to the conclusion that although large cracks will most likely be instrumental in causing an earlier breakdown in passivity of the reinforcing steel and thus will be responsible for initiating the corrosion process earlier, there is no logical explanation why the rate of corrosion would be influenced by the surface cracks unless these cracks represent widespread internal cracking.

Mehta and Gerwick[9] emphasized the role of concrete microstructure in the cracking–corrosion–cracking cycle (Fig. 5.7). Since microcracks already exist in concrete, particularly in the transition zone, the authors proposed that the widening of microcracks due to any number of possible causes creates an interconnected network of cracks which increases the permeability of concrete. In the marine environment, an increase in the permeability of concrete probably occurs under service conditions. The increase in permeability facilitates all of the conditions that are necessary for the corrosion process to proceed, namely the depassivation of the steel surface by chloride ions or by carbonation, the lowering of the electrical resistivity of concrete by almost complete saturation, and the availability of enough dissolved oxygen at the steel surface. Obviously, wide cracks in the cover of a reinforced concrete element will, in the beginning, have no effect on corrosion of steel because, before any significant

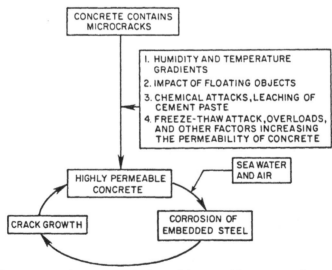

Fig. 5.7. Diagrammatic representation of the cracking–corrosion–cracking cycles in concrete (reproduced, with permission, from Ref. 9).

corrosion can take place, an overall increase in the permeability of the concrete must proceed through the propagation of internal microcracks. In the long run, however, once the microcracks interconnect and the permeability increases, the large cracks in the cover will behave like mighty rivers, with their flow direction reversed. Instead of channelling the freshwater flow from the hinterland tributaries into the sea, the rivers (large cracks) will carry the seawater flow via small tributaries (microcracks) into the hinterland (interior of concrete). When there is a plentiful supply of chloride, water, and oxygen in the cement paste–steel transition zone, the stage is set for the corrosion–cracking–corrosion cycles leading to serious deterioration of both reinforcing steel and concrete.

To prevent the reinforcement from corroding, the use of the cathodic protection method (see Chapter 8), epoxy-coated rebars, and calcium nitrite as an admixture in fresh concrete are among the several possibilities. The use of low-permeability concrete mixtures and the maintenance of low permeability in service however offers the best and the most effective approach.

FROST ACTION

In cold climates, frost action is a frequent cause for cracking and spalling of unprotected concrete (i.e. concrete without adequate air entrainment). In many coastal areas the air temperature drops considerably below the freezing point of water. However, the presence of warm currents prevents the water below the ocean surface from freezing. The twice-daily tidal action thus exposes the concrete in the splashing zone to two cycles a day of freezing and thawing, which amounts to over two hundred cycles every year in many locations. In cold climates, therefore, it is not coincidental that the greatest structural damage due to frost action occurs in the upper region of the splashing zone which is exposed to the largest number of freezing and thawing cycles.

The deterioration of concrete by frost action (i.e. exposure to freezing and thawing cycles), is generally attributed to the complex microstructure of the hydrated cement paste in concrete. Powers'[10] description of the mechanism of expansion of a saturated cement paste on freezing, and the mechanism by which air entrainment is able to reduce the expansion, is as follows:

> When water begins to freeze in a capillary cavity, the increase in volume accompanying the freezing of the water requires a dilation of the cavity equal to 9% of the volume of frozen water, or the forcing of the amount of excess water out through the boundaries of the specimen, or some of both effects. During this process, *hydraulic pressure* is generated and the magnitude of that pressure depends on the distance to an 'escape boundary', the permeability of the intervening material, and the rate at which ice is formed. Experience shows that disruptive pressures will be developed in a saturated specimen of paste unless every capillary cavity in the paste is not farther than three or four thousandths of an inch from the nearest escape boundary. Such closely spaced boundaries are provided by the correct use of a suitable air-entraining agent.

Some researchers believe that in porous solids the primary cause of expansion on freezing involves the large-scale migration of water from small pores to large capillary cavities. According to Litvan,[11] the water held in very small pores of the hydrated cement paste does not freeze at the normal freezing point of water in large capillaries. This is because the formation of ice crystals requires a rearrangement of water molecules; since the mobility of oriented water molecules in small pores is rather limited, it continues to exist in a supercooled state without freezing. This creates a thermodynamic disequilibrium between the frozen water in large capillaries (which is in a low-energy state), and the supercooled water in small pores (which is in a high-energy state). The difference in entropy between ice and supercooled water forces the latter to migrate toward the low-energy sites (large capillaries). Since capillaries are already filled with ice, they are unable to accommodate any more water. This situation causes hydraulic pressures in the system, which can eventually lead to expansion and cracking. The extent of mechanical damage is determined by the degree of saturation, the temperature gradient, and the permeability of the system. Powers[10] found a decreasing tendency for saturated cement pastes to expand, with increasing entrainment of air in the form of 0·05–1 mm size bubbles, uniformly distributed in the cement paste (such that the void spacing was within 0·1 to 0·2 mm).

In addition to adequate and proper air entrainment, an important controlling factor, which greatly influences the frost resistance, is the permeability of concrete. A well-cured concrete with low water/cement ratio and high cement content, may start out as unsaturated and impermeable. However, microcracking from any cause, such as normal thermal and humidity cycles in service, can increase the permeability and subsequently the degree of saturation. At this stage, further deterioration due to frost action can be reduced by the presence of properly entrained air. Otherwise, frost action will lead to more microcracking. A permeable, weakened concrete, then falls an easy victim to other physical and chemical agents of destruction, such as sulfate attack, alkali–silica expansion, and corrosion of the reinforcing steel.

The results from recent experimental studies by Moukwa[12] on

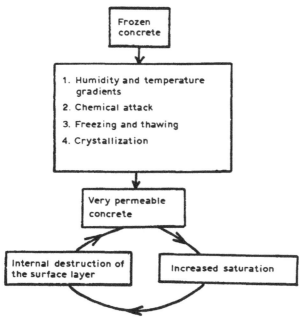

Fig. 5.8. Diagrammatic representation of the mechanisms of disruption of concrete exposed to thermal shock from cycles of freezing and thawing (reproduced, with permission, from Ref. 12).

concrete exposed to cold water in the tidal zone have confirmed that thermal shocks from cycles of wetting–drying and freezing–thawing produce microcracks in concrete and allow further saturation, which destroys the surface thus making concrete more permeable and more vulnerable to various types of physicochemical attacks. In Fig. 5.8 the author has presented a conceptual representation of the mechanism of concrete deterioration. It may be noted that the diagrammatic representation of Moukwa's hypothesis is very similar to Mehta and Gerwick's hypothesis (Fig. 5.7) on the cracking–corrosion–cracking cycles in reinforced concrete structures in the marine environment.

CHEMICAL ATTACKS

A comprehensive review of the possible patterns of decomposition of hydrated portland cement paste, as shown in Fig. 5.3, has been presented by the author.[13] The surface of concrete is the first line of defence against seawater. With a high-quality or an impermeable concrete skin, the chemical attack by seawater can be limited essentially to the surface. For a variety of reasons already discussed, should a concrete become permeable and permit the penetration of seawater, the door will open for several harmful reactions, which occur between seawater and the constituents of hydrated cement paste. It should never be forgotten that the seeds for potential decomposition of cementitious constituents of concrete are already present in the situation because hydrated portland cement paste is essentially alkaline (it consists of compounds of calcium that are in contact with a pore fluid of >13 pH), whereas seawater contains acidic ions and has a pH which is close to neutral (8·2–8·4).

Since aragonite ($CaCO_3$) is the most stable mineral in the seawater environment, the outermost zone of concrete near the surface is usually found to contain aragonite. The solid hydration products in a well-hydrated cement paste, made with a normal ASTM Type I portland cement, usually consist of calcium hydroxide, calcium monosulfoaluminate hydrate, and calcium silicate hydrates. All three are susceptible to decomposition by CO_2 attack (Table 5.1). With the low CO_2 concentrations normally present in seawater, the consequences of the CO_2 attack are not serious because the formation of insoluble aragonite reduces the permeability of concrete. However, this type of protection against further chemical attack will not be available in the splashing zone where the wave action can remove the chemical interaction products as soon as they are formed. Also, in estuary or bay waters with a high content of dissolved CO_2, aragonite is transformed to calcium bicarbonate which, being soluble, is leached away. This has the effect of increasing the porosity of concrete and therefore reducing the strength.

As shown in Table 5.1, magnesium salts present in seawater can also enter into deleterious chemical reactions with portland cement paste, resulting in the formation of brucite (magnesium hydroxide) and soluble products, such as calcium chloride and calcium sulfate. In old concretes, as a result of ion exchange between seawater and calcium silicate hydrates present in the hydrated portland cement paste, a magnesium silicate ($4MgO \cdot SiO_2 \cdot 8H_2O$) has been identified. According to Regourd,[14] the substitution of the magnesium silicate hydrates for calcium silicates hydrates makes the concrete weak and brittle.

It may be noted that the concentration of $MgSO_4$ in seawater (approximately 2200 g/liter) is large enough for *sulfate attack,* which is normally associated with expansion and loss of mass resulting from the formation of ettringite and gypsum, respectively.[1] Although many investigators, including Mehta and Haynes[15] found both gypsum and ettringite in concrete deteriorated by the action of seawater, the expansion and cracking normally associated with ettringite formation does not occur in the case of seawater attack. This is attributable to the formation of ettringite in a chloride environment, rather than in an alkaline environment, which is essential for the swelling of ettringite by water adsorption.[1]

The chloride ions from seawater tend to penetrate deep into the interior of concrete. Although chloride can react with high-C_3A cement pastes to form calcium chloroaluminate hydrate, the reaction product is unstable in the presence of sulfate ions and CO_2. This is why the stable products of interaction between

Table 5.1. Decomposing Action of Seawater on the Constituents of Hydrated Portland Cement.

Seawater component which can enter into deleterious chemical reactions with hydrated portland cement	Possible chemical reactions	Physical effects associated with chemical reactions
Carbon dioxide Small quantities of dissolved CO_2, derived mainly from absorption of atmospheric CO_2, always present in seawater. However, decaying vegetable matter can lead to substantially larger and harmful concentrations of dissolved CO_2, which are generally reflected by reduction of the seawater pH to values less than 8	$CO_2 + Ca(OH)_2 \xrightarrow[\text{aragonite}]{CO_2} \underset{\text{bicarbonate of calcium}}{Ca(HCO_3)_2}$ $CO_2 + [Ca(OH)_2 + 3CaO \cdot Al_2O_3 \cdot CaSO_4 \cdot 18H_2O] \rightarrow 3CaO \cdot Al_2O_3 \cdot CaCO_3 \cdot xH_2O + \underset{\text{gypsum}}{CaSO_4 \cdot 2H_2O}$ $3CO_2 + 3CaO \cdot 2SiO_2 \cdot 3H_2O \rightarrow \underset{\text{aragonite}}{3CaCO_3} + 2SiO_2 \cdot H_2O$	Both calcium bicarbonate and gypsum are soluble in seawater. Loss of material and weakening or mushiness of hardened cement paste can therefore be associated with the formation of these compounds. Since all the hydration products of portland cement, including the calcium silicate hydrate, can be decomposed by carbonation reactions, permeable concretes in seawater containing larger than normal CO_2 concentration are likely to deteriorate
Magnesium salts Typically, sea water contains 3200 ppm $MgCl_2$ and 2200 ppm $MgSO_4$. Regarding cement hydration products, these magnesium salts, even in the small concentrations present, are considered harmful	$MgCl_2 + Ca(OH)_2 \rightarrow \underset{\text{brucite}}{Mg(OH)_2} + CaCl_2$ $MgSO_4 + Ca(OH)_2 \rightarrow Mg(OH)_2 + \underset{\text{gypsum}}{CaSO_4 \cdot 2H_2O}$ $MgSO_4 + [Ca(OH)_2 + 3CaO \cdot Al_2O_3 \cdot CaSO_4 \cdot 18H_2O] \rightarrow Mg(OH)_2 + \underset{\text{ettringite}}{3CaO \cdot Al_2O_3 \cdot 3CaSO_4 \cdot 32H_2O}$ $MgSO_4 + [Ca(OH)_2 + 3CaO \cdot 2SiO_2 \cdot 3H_2O] \rightarrow 4MgO \cdot SiO_2 \cdot 8H_2O + CaSO_4 \cdot 2H_2O$	$CaCl_2$ and gypsum, being soluble in seawater, lead to material loss and weakening. Formation of ettringite is associated with expansion and cracking. It is reported that the conversion of $3CaO \cdot 2SiO_2 \cdot 3H_2O$ to $4MgO \cdot SiO_2 \cdot 8H_2O$ is associated with brittleness and strength loss

Reproduced, with permission, from Ref. 13.

portland cement paste and CO_2, Mg^{2+}, and SO_4^{2-} can be detected in the outer zone of concrete, whereas the chloroaluminate hydrate has not been detected. The formation of chloroaluminate hydrate owing to seawater attack on portland cement has not been reported to be harmful to strength. However, there would be a general loss of strength and elastic properties when the paste becomes saturated with chloride environment instead of the hydroxyl-ion environment.

The chemical attack of seawater is generally confined to the cement paste. However, in some instances, when *aggregates containing reactive silica* are used for making concrete, the attack may also involve the aggregate particles. The Na^+ and the chloride ions present in seawater are known to accelerate the alkali–silica reaction which, under certain circumstances, can lead to expansion and cracking of concrete.

The mechanisms of expansion of concrete due to the alkali-aggregate reaction are rather complex; however, the following simple explanation by Figg[2] seems adequately to cover the aggregates containing reactive silica:

> Alkali–silica reaction involves formation of an alkali silicate gel by fission of siloxy groups in siliceous aggregates by hydroxyl ions derived from portland cement followed by combination with sodium (or potassium) ions, present in cement paste pore fluid. The gel attracts water molecules, which causes a hydraulic pressure of the order of 10–15 MPa, well above the tensile strength of conventional concretes.

According to Idorn,[16] only a few rock types can be considered non-reactive with certainty; these include carbonate rocks (without magnesium or silica), basalts, gneiss, and granite. When igneous, crystalline, or volcanic rocks are exposed to severe metamorphic alterations or to long-time weathering in tropical heat, they also become susceptible to alkali attack. Figure 5.9 illustrates the mineral types which under various conditions have been found to cause the deleterious alkali–silica reaction. When affected by the expansive alkali–silica reaction, characteristic features of microcracking both within and radiating from siliceous aggregate particles of different mineralogy and morphology, are shown in Fig. 5.10. Since temperature, humidity, and salinity promote the alkali–silica reaction, the seawater environment can

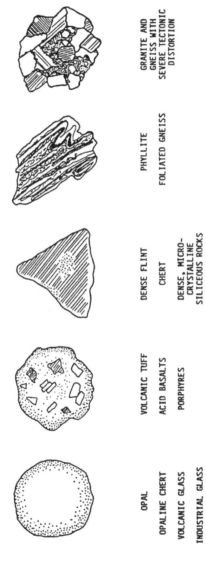

Fig. 5.9. Sketches displaying texture, morphology and compositions of the primary alkali susceptible rock types (reproduced, with permission, from Ref. 16).

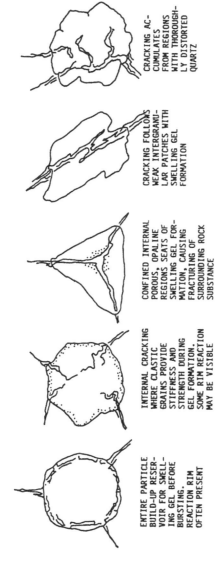

Fig. 5.10. Characteristic features of microcracking within and radiating from siliceous aggregate particles of different mineralogy and morphology, when affected by expansive alkali–silica reactions (reproduced, with permission, from Ref. 16).

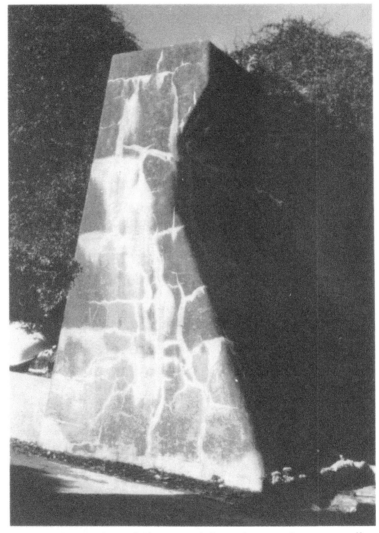

Fig. 5.11. Cracking and exudation on reinforced concrete wave-wall, a combination of alkali–silica reaction and wetting–drying cycles (courtesy: J. Figg).

be an activating factor especially in tropical regions. A photograph of typical cracking as a consequence of alkali–silica reaction, and cycles of wetting and drying is shown in Fig. 5.11.

From his extensive experience, Idorn says that it is not necessary always for the alkali–silica reaction to result in deleterious expansion. As a consequence of the alkali–silica reaction *unreinforced concrete elements* may occasionally undergo sufficient deterioration to call attention to the structural soundness, especially with marine structures in a hot climate. In contrast, with properly *reinforced elements,* the stresses generated by the alkali–silica reaction alone may not be large enough to cause a significant decline in structural performance. However, should microcracking result from expansion associated with the alkali–silica reaction, it can play an important part in the corrosion of reinforcing steel. Therefore, considering the overall implications of the alkali–aggregate attack in marine structures, the phenomenon cannot be entirely ignored.

The commonly used methods to control the alkali–silica reaction in concrete include the use of low-alkali portland cement, and avoiding the use of reactive aggregates and alkali-contaminated aggregates. The use of a portland–pozzolan cement or a portland–blast-furnace slag cement or corresponding mineral admixtures, with proven ability to reduce the expansion in a reference cement–aggregate mixture, is also an accepted method of dealing with the problem when other solutions are not feasible. Finally, it should be noted that the mechanism of expansion (swelling of alkali silicate gel) requires the availability of water. Therefore, in spite of alkali–aggregate activity, deleterious expansion is unlikely when the permeability of concrete is low enough to prevent water saturation.

Chemical attack on concrete as a result of *bacterial activity* has long been observed in sewage pipes. Many researchers including Sullivan and Khoury[17] have reported that there is evidence of similar bacterial activity from hot oil (40–50°C) in the concrete oil storage tanks in the North Sea. The oil serves as a food source for a variety of bacteria which grow under warm acidic conditions. The three main products of aerobic and anaerobic bacterial activity in hot oil–seawater environment, namely acetic acid, H_2S, and sulfuric acid, are harmful to concrete. The acids attack the alkaline constituents of the hydrated cement paste and convert them into soluble products which are leached away. Sulfur-oxidizing bacteria (SOB) are the source of sulfuric acid which is formed by bacterial action on H_2S. Sulfuric acid is highly corrosive to both concrete and reinforcing steel.

The rate of acid attack on concrete in oil storage tanks is

influenced by a number of factors. According to Sullivan and Khoury[17] the impermeable layer of aragonite ($CaCO_3$) and brucite on the concrete surface offers some protection. However, this protection may be lost due to thermal cycling resulting from entry and removal of hot crude oil which causes microcracking in the concrete surface. The elevated temperatures enhance both the bacterial activity and chemical attack on concrete, as discussed below.

CRYSTALLIZATION PRESSURE OF SALTS

In porous materials, the crystallization pressure of salts from supersaturated solutions is known to produce stresses that may be large enough to cause cracking and spalling. Such situations can arise in tunnel liners, slabs, and walls of *permeable* concrete, when one side of the structural element is in contact with salt water (seawater) and the other sides are exposed to warm air (i.e. subject to evaporation). In general, the higher the degree of supersaturation, the greater the crystallization pressure. Even with as low a degree of supersaturation as 2, crystallization of NaCl can exert pressures of the order of 600 atmospheres, and the resulting stress is large enough to break most rocks.[1]

According to Figg,[2] saline water is able to creep upwards in concrete by capillarity so that in the evaporation zone, usually 0·3–0·5 m above the water level, salt crystallization causes concrete spalling which looks similar to spalling by frost action. Thus, in the *splashing zone,* deterioration of concrete may partly result from salt crystallization pressure. Salt crystallization damage to the surface of horizontal concrete slabs can also happen, where puddles are permitted to form (Fig. 5.12). Successive cycles of salt crystallization are also known to result in onion-skin type surface spalling, called 'Liesegang rings', which have been noted with weathering of certain rocks (Fig. 5.13).

The quality of skincrete (see Chapter 7) plays an important part in determining the rate at which salt scaling occurs on the surface of a concrete structure exposed to the marine environment.

Fig. 5.12. Concrete dock surface spalled and eroded by salt crystallization from evaporated seawater puddles (courtesy: J. Figg).

Fig. 5.13. Surface erosion of concrete slab exposed to seawater splashing where evaporation has produced Liesegang-ring 'onion-skin' spalling due to successive salt crystallization cycles (Courtesy: J. Figg).

Therefore, in good concreting practice a smooth and impermeable surface finish must be given a top priority. This is particularly important with modern concrete mixtures containing silica fume which tend to be sticky and hard to finish.

ATTACK BY MICRO-ORGANISM

The presence of sulfur from any source in an aerobic environment promotes the growth of micro-organisms known as the Thiobacilli. Thiobacilli, a diverse group of bacteria, are able to oxidize sulfur to sulfate, with eventual formation of sulfuric acid which is highly corrosive to both steel and concrete. Kulpa and Baker[18] described three classes of Thiobacilli that are involved in the microbial deterioration of concrete. The data in Table 5.2 show the pH ranges of growth and typical food (energy) sources of each class of the bacteria. Thus concrete pipes and other structures containing oil may be susceptible to attack from Thiobacilli. According to the authors, besides sulfur only oxygen and carbon dioxide are needed for growth of this type of bacteria.

Conceivably, the first phase of aerobic oxidation of sulfur in the environment close to concrete would involve a neutral or slightly acidic pH and the presence of *Thiobacillus neapolitanus* (Table 5.2). With the oxidation of sulfur, as the pH is lowered, a second group of organisms characterized by *Thiobacillus thiooxidans* becomes active. These organisms are vigorous sulfur oxidizers and are capable of lowering the pH of the environment to $1 \cdot 0$ or lower. A continuing source of sulfur supply (viz. through flowing oil or

Table 5.2. Different Classes of Thiobacilli.

Representative species	pH growth range	Energy from oxidation of
T. neapolitanus	7·0–4·0	$S°$, $S_2O_3^{-2}$, H_2S
T. ferrooxidans	5·8–1·3	$S°$, H_2S, Fe^{+2}
T. thiooxidans	4·5–<1·0	$S°$, $S_2O_3^{-2}$, H_2S

Reproduced, with permission, from Ref. 18.

Table 5.3. Proposed Sequence of Events for Microbial Deterioration of Concrete by Sulfur-Oxidizing Bacteria.

1. Presence of a sulfur source
   a. S in environment
   b. H_2S in environment
2. Activity by moderate, neutral pH sulfur-oxidizing bacteria. Results in pH of environment changing to a pH of 4–5
3. Continued sulfur presence
4. pH reaches approximately 4·5 allowing growth of acidophilic Thiobacilli, (*T. thiooxidans* and *T. ferrooxidans*)
5. pH drops to 2–3, significant H_2SO_4 production occurs, concrete deterioration mediated by sulfuric acid occurs and continues as long as sulfur is present

Reproduced, with permission, from Ref. 18.

sewage) is essential to support the growth of these bacteria. A third organism, *Thiobacillus ferrooxidans* (Table 5.2) has the unique ability to oxidize either sulfur or iron and can contribute directly to the destruction of reinforcing steel. Table 5.3 shows the proposed sequence of processes involved in the microbial deterioration of concrete by sulfur–oxidizing bacteria.

Interestingly, in the presence of sulfate and anaerobic conditions that may be found in some sewage pipes, there is another bacterial group (Desulfovibrio) which can reduce sulfate to hydrogen sulfide. As described above, H_2S can then become the source of energy for growth of Thiobacilli. Fortunately, owing to the high pH of hydrated portland cement, dense and impermeable concretes should remain immune to such bacterial attacks. Only when there is a substantial loss of alkalinity by leaching of the cement paste, would a concrete become prone to attack by microorganisms.

CONCLUDING REMARKS

From the above review of the most common causes of concrete deterioration in seawater it appears that the same mechanism may be behind many of the deleterious phenomena. For instance, in frost action, sulfate attack, alkali–aggregate attack, or corrosion of reinforcing steel, some researchers believe that the expansion of concretes is primarily due to the development of internal hydraulic pressures. Consequently, in every case the *degree of water saturation of concrete,* which is dependent on the permeability, will play an important role. Experience shows that even relatively impermeable concrete mixtures can gradually become permeable in service, and thus vulnerable to seawater attack. There is an increasing awareness that thermal stresses resulting from improper cooling of steam-cured products or from rich concrete mixtures with high heat of hydration, are often a neglected source of microcracking, which is the principal cause for enhancement of the permeability of concrete in structures. Selection of materials, mix proportions, and concreting practice—which is the subject of the next two chapters—must be done with extra care to reduce microcracking when a structure is exposed to temperature extremes, i.e. frequent cycles of freezing and thawing or heating and cooling.

REFERENCES

1. Mehta, P. K., *Concrete: Structure, Properties, and Materials,* Prentice-Hall, Englewood Cliffs, NJ, 1986, pp. 105–69.
2. Figg, J., Salt, sulfate, rust and other chemical effects, *Proceedings Gerwick Symposium on Durability of Concrete in Marine Environment,* ed. P. K. Mehta, Dept. of Civil Engineering, University of California at Berkeley, 1989, pp. 50–68.
3. Mehta, P. K., Durability of concrete in marine environment—an overview, *Proceedings Gerwick Symposium on Durability of Concrete in Marine Environment,* ed. P. K. Mehta, Dept. of Civil Engineering, University of California at Berkeley, 1989, p. 70.
4. Gjørv, O. E., Steel corrosion in marine concrete structure—an overview, *Proceedings Gerwick Symposium on Durability of Concrete in Marine Environment,* ed. P. K. Mehta, Dept. of Civil Engineering, University of California at Berkeley, 1989, pp. 77–89.
5. Gjørv, O. E., Vennesland, O. & El-Busaidy, A. H. S., Diffusion of dissolved oxygen through concrete, *Materials Performance,* **25**(12) (1986), pp. 38–44.

6. Gjørv, O. E., Vennesland, O. & El-Busaidy, A. H. S., Electrical resistivity of concrete in the oceans, *Proceedings Offshore Technology Conference*, Houston, Texas, 1977, pp. 581–8.

7. Vennesland, O. & Gjørv, O. E., Silica fume concrete—protection against corrosion of embedded steel, *Fly Ash, Silica Fume, Slag and Other Mineral By-Products in Concrete*. ed. V. M. Malhotra, ACI SP-79, 1983, pp. 719–29.

8. Beeby, A. W., Cracking, cover, and corrosion of reinforcement, *Concrete International*, **5**(2) (1983), pp. 35–40.

9. Mehta, P. K. & Gerwick, B. C., Cracking-corrosion interaction in concrete exposed to marine environment, *Concrete International*, **5**(10) (1982), pp. 45–51.

10. Powers, T. C., The physical structure and engineering properties of concrete, *Bulletin 90*, Portland Cement Association, Skokie, IL, 1958.

11. Litvan, G. G., Frost action in cement in the presence of de-icers, *Cement and Concrete Research*, **6** (1976), 351–6.

12. Moukwa, M., Deterioration of concrete in cold seawaters, *Cement and Concrete Research*, **20** (1990), 439–46.

13. Mehta, P. K., Durability of concrete in marine environment—a review. *Performances of Concrete in Marine Environment*, ed. V. M. Malhotra, ACI SP-65, 1980, pp. 1–20.

14. Regourd, M., The action of seawater on cements, *Annales de l'Institut Technique du Batiment et des Travaux Publics*, No. 329 (1975), 86–102.

15. Mehta, P. K. & Haynes, H., Durability of concrete in seawater, *J. ASCE Structural Division*, **101**(ST8) (1975), 1679–86.

16. Idorn, G. M., Marine concrete technology—viewed with Danish eyes, *Proceedings Gerwick Symposium on Durability of Concrete in Marine Environment*, ed. P. K. Mehta, Dept. of Civil Engineering, University of California at Berkeley, 1989, pp. 34–7.

17. Sullivan, P. J. E. & Khoury, G. A., The effect of bacterial activity on North Sea crude oil and concrete, *Proceedings Sixth International Offshore Mechanics and Arctic Engineering Symposium*, American Society of Mechanical Engineers, 1987.

18. Kulpa, C. F. & Baker, C. J., Involvement of sulfur-oxidizing bacteria in concrete deterioration, *Proceedings, Paul Klieger Symposium on Performance of Concrete*, ed. D. Whiting, ACI SP-122, 1990, pp. 313–322.

Chapter 6
Selection of Materials and Proportions for Durable Concrete Mixtures

From the performance of concrete in the marine environment (Chapter 4) and a review of the principal causes of concrete deterioration (Chapter 5), it is clear that the *permeability of concrete* is the most important factor determining the long-time durability. Therefore, with any new construction not only is it important to select materials and proportions for the concrete mixture that are most likely to produce a low-permeability product on curing but also necessary to maintain the watertightness of the structure as long as possible through the intended service life. In short, careful attention should be paid to *all three* of the following aspects of concrete construction:

- selection of concrete-making materials and mix proportions
- good concreting practice
- measures to prevent the widening of pre–existing microcracks in concrete during service.

It should be noted that many of the recent concrete sea structures, built during the last 15–20 years, are required to withstand unprecedented stress conditions. For example, coastal and offshore structures in the North Sea and the Arctic are exposed to enormous hydrostatic pressures, impact loading, frost action, and abrasion/erosion loss from floating ice. Coastal structures in the Middle East are exposed to numerous cycles of temperature extremes (i.e. hot days and cold nights). Consequently, these structures made with high-strength concrete are very heavily reinforced as well as prestressed. The protection of the embedded steel with a low permeability concrete is of paramount importance from the standpoint of durability. To meet this challenge, i.e. to produce high-strength and low-permeability concrete, many new materials have been developed for use during the last two decades.

The technology for producing high-strength and high-impermeability concrete is therefore already available, and has been incorporated into field practice and industry codes. It is too early to pass any definitive judgment on this technology although most results are optimistic. For example, no evidence of impairment in durability has been reported for the high-strength and high-impermeability concrete used for the construction of North Sea offshore concrete platforms, some of which are now 20 years old. On the other hand, signs of deterioration are reported in concretes from the Manma-Sitra Causeway Bridge in Bahrain, which is only 15 years old.

In regard to the industry codes, it is interesting to note that the American Concrete Institute has no special codes for conventional coastal marine structures. The *ACI Manual of Concrete Practice* contains numerous guidelines for general structures (such as buildings and bridges, parking garages, silos and bins, chimneys and cooling towers, and nuclear and sanitary structures) which are followed for the design of conventional marine structures. ACI Committee 357 has prepared a recommended practice for the design of *offshore* structures, a state-of-the-art report on Arctic offshore structures and another report on barge-like structures. This committee is now engaged in the process of preparing an *ACI Guide for Design of Concrete Marine Structures*.

The recommended practice and the industry codes for concrete for offshore structures tend to be more stringent than for conventional coastal structures. From the standpoint of concrete durability, therefore, the quality requirements for offshore concrete mixtures are generally applicable to all marine structures. Although a new standard for marine concrete structures is being developed in Canada, in the meantime, both in the United States and Canada, the ACI Committee 357 recommendations are generally followed for the construction of fixed reinforced and prestressed concrete structures in the marine environment. Builders in Europe and Asia prefer to follow other codes, such as the British Standard Specifications, the UK Department of Energy Code, and the FIP (International Federation of Prestressed Concrete Structures) Code of Practice. A comparison of many standards shows that most of them are not significantly different from the ACI or FIP codes. Also, it should be noted that although both the ACI and FIP recommendations for design and construction of concrete offshore structures, which are reviewed in this chapter, were revised in 1985 (they are revised periodically), they tend to lag behind the latest North Sea concrete practice. For long-term durability of concrete, therefore, it will be useful to compare the current North Sea offshore concrete technology with the ACI and FIP recommendations for the selection of concrete-making materials, mixture proportions, and construction practice for offshore structures.

SELECTION OF CONCRETE-MAKING MATERIALS

Until about the middle of this century, concrete mixtures were generally composed of four components: cement, water, fine aggregate, and coarse aggregate. Most of the concrete produced today is a multicomponent product containing one or more admixtures in addition to the four basic components. Furthermore, for every component the producer usually has several choices that could influence the cost of the end product and its behavior in service. Therefore, before a discussion on mix proportioning, it is useful to acquire a knowledge of the choices available in cements, aggregates, and admixtures that are described in Chapter 3. Materials selection is usually an art because there are always some undesirable effects associated with the use of a given material. Critical evaluation of the industry codes and current recommended practice is the first necessary step towards acquisition of this art. With a multicomponent mixture, mutual compatibility problems and interaction effects cannot be easily predicted; hence laboratory trial batches and field experience are also necessary to arrive at optimal solutions.

Cements

As described in Chapter 3, the ASTM C 150 Standard Specification for Portland Cement covers eight types of portland cements. However, only the following four types are generally produced in the United States and other countries of the world, most of which have their own specifications with minor differences:

1. *General-purpose, normal portland cement (ASTM Type I)*: The content of highly active components, C_3S and C_3A is approximately 45–55%, and 8–12%, respectively, and the fineness is typically 350–400 m^2/kg Blaine. No limits are prescribed for the compound composition.
2. *General-purpose, modified portland cement (ASTM Type II), with moderate sulfate resistance and moderate heat of hydration*: The amounts of C_3S and C_3A range from 40% to 45% and from 5% to 7%, respectively, and the fineness is usually 300 m^2/kg Blaine. The ASTM C 150 Specification limits are given in Chapter 3.
3. *High early strength portland cement (ASTM Type III)*: Also called rapid hardening cement, this is specified when a higher than normal rate of strength development at early ages (viz. 1-d, 3-d, 7-d) is desired at moderate curing temperatures, or a normal rate of strength development is desired under colder ambient conditions (e.g., 5–10°C). High early strength is due to larger amounts of C_3S and C_3A than are normally present in ASTM Type I portland cement, and higher fineness (typically 500 m^2/kg). The high heat of hydration associated with this cement can be a major disadvantage under certain conditions. It should be noted that, at a given water to cement ratio, a concrete mixture made with ASTM Type III portland cement may have a similar or even somewhat lower *ultimate strength* when compared to a corresponding concrete mixture made with

ASTM Type I or Type II portland cement. Thus ASTM Type III cement is not necessarily suitable for making high-strength concrete, although it is suitable for making high early strength concrete.

4. *Sulfate-resistant portland cement (ASTM Type V)*: For sulfate waters containing more than 1·5 g/liter and up to 10 g/liter sulfate ions, portland cements with less than 5% C_3A are specified to obtain sulfate resistance. The C_3S content and the fineness of ASTM Type V cements are usually similar to the ASTM Type II cement, therefore both the strength development rate and the heat of hydration are lower than the ASTM Type I portland cement.

In addition to portland cement, blended portland cements, such as Type IS cement containing 30% to 65% rapidly cooled blast-furnace slag or Type IP cement containing 15% to 30% pozzolan (see Chapter 3) are also widely used throughout the world. Compared to ordinary portland cements, the blended portland cements are characterized by lower heat of hydration, lower rate of strength development, and better chemical resistance. However, it should be noted that these characteristics from the use of blended portland cements can also be obtained by incorporating a pozzolan or a ground granulated blast-furnace slag as a mineral admixture into portland cement concrete mixtures.

The ACI and FIP recommendations and the North Sea field practice for concrete-making materials are summarized in Table 6.1. It may be noted that the ACI 357R-84 recommendations for the design of offshore concrete sea structures suggest the use of ordinary portland cement with C_3A content between 4% and 10%. Similarly, the corresponding FIP recommendations also suggest the use of moderate C_3A portland cements especially in the splashing zone and the atmospheric zone. The maximum C_3A limit is probably based on consideration of sulfate attack, because the high sulfate concentration of seawater (approximately 2·5 g/liter) normally would have required the use of ASTM Type V portland cement ($<5\%$ C_3A). In Chapter 4, a review of several case histories of long-time seawater exposure of concretes containing 14–17% C_3A portland cements, however, showed no evidence of expansive cracking due to ettringite formation

Table 6.1. Concrete-Making Materials for Offshore Structures.

Material	ACI 357R-84 Recommendation	FIP-1985 Recommendation	North Sea field practice
Cement composition	C_3A content of portland cement should not be less than 4% to provide protection for the reinforcement. The maximum C_3A should be limited to 10% to obtain resistance to sulfate attack	In the splash zone and atmospheric zone, portland cements with moderate C_3A content are recommended. Rapid-hardening cements should only be used for repair. Low heat of hydration cements are preferred for structures with heavy dimensions	For Gullfaks C (1986–87), a finer-ground portland cement (400 m^2/kg Blaine), conforming to ASTM Type II composition (49% C_3S, 5·5% C_3A) is used. Concrete for Beryl A (1974) was made with a typical ASTM Type I cement (55% C_3S, 8% C_3A, 300 m^2/kg Blaine)
Admixtures Chlorides	No chlorides should intentionally be added. Total water soluble chloride ion of a concrete mixture from all the component materials should not exceed 0·1% by weight of cement for normal reinforced concrete, and 0·06% prestressed concrete	$CaCl_2$ or admixtures containing more than 0·1% chloride by weight of cement should not be used	No chloride admixtures are used
Chemical and air entrain-ing	Where freeze–thaw durability is required, the concrete should contain entrained air as recommended by ACI 201.2R	Air entraining agents, workability aids, and retarders are often essential to obtain optimum mix design, but precautions should be taken to evaluate the side effects of each admixture type before use	For Gullfaks C, air entrainment was used for concrete in the splash zone. A spacing factor of 0·25 mm was obtained with 3–5% air. The concrete also contains 6 liters/m^3 of a naphthalene sulfonate type superplasticizer
Pozzolanic	Pozzolans conforming to ASTM C618 (only natural pozzolans and fly ash are covered) may be used provided tests are made to ascertain their relative advantages and disadvantages specially in regard to sulfate resistance, workability, and corrosion of steel	High-quality pozzolanic materials, such as special silica fumes may be added to produce improved strength, durability, and workability	Gullfaks C concrete for the splash zone contains approximately 5% silica fume by weight of cement
Aggregates	Natural sand and gravel, or crushed rock conforming to ASTM C33, and lightweight aggregate conforming to ASTM C330. Marine aggregates may be used provided they have been washed to meet the chloride ions limits. No limits on maximum aggregate size are given	Aggregates likely to undergo physical or chemical changes in concrete to be avoided. Marine aggregates should not be used unless the chloride content is at an acceptable level, and unless the aggregates have a sufficiently low seashell content	High-quality natural sand and gravel are used. Concrete in older structures contained up to 32 mm coarse aggregate, but in recently built (1981–87) offshore structures contains 20 mm maximum aggregate size

A rigid adherence to the 10% maximum C_3A content, as specified by ACI 357R, therefore does not seem to be justified in the case of concrete mixtures for marine structures.

A similar situation exists in regard to the lower C_3A limit (4% minimum), which is based on the results of a rather limited study reported by Verbeck.[1] The author found that test piles of reinforced concrete made with high-C_3A portland cements (8–12% C_3A) showed less corrosion-cracking in seawater at a test site in St Augustine, Florida, than similar piles made with low-C_3A portland cements (2–5% C_3A). This was attributed to possible insolubilization of chloride ions in the form of calcium chloroaluminate hydrate, $C_3A \cdot CaCl_2 \cdot 11H_2O$. It was pointed out by Mehta[2] that this sort of chloride removal probably does not take place in the seawater environment, as the presence of the chloroaluminate has seldom been reported from deteriorated concrete exposed to seawater. This is because the compound is unstable in the presence of sulfate and CO_2. The tendency for the calcium chloroaluminate hydrate to decompose and provide chloride ions to the solution has also been reported by Rosenberg and Gaidis,[3] who believe that the problem of chloride corrosion in reinforced concrete cannot be mitigated by the use of high-C_3A portland cements.

In the North Sea field practice (Table 6.1),[4] a coarse-ground (300 m^2/kg), Type I portland cement (55% C_3S, 5·5% C_3A) was used for the construction of Beryl A (1975). More recently, in 1986–87, a finely ground (400 m^2/kg) Type II portland cement (49% C_3S, 8% C_3A) was used for the construction of Gullfaks C. It is stated that the choice of cement is governed by 'constructibility' considerations. For instance, the setting and hardening characteristics of concrete for slip-formed elements, and cement-admixture compatibility (i.e. workability, slump loss, air-entrainment, etc.) are the issues which demand immediate attention with regard to the cement selection at site. Minor concerns regarding the effect of cement composition on concrete durability can be disregarded if suitable steps are taken to insure the watertightness of concrete.

Considering some of the properties of concrete, such as workability, response to microcracking due to thermal stresses, and diffusion rate of chloride ions, experience shows that most ASTM Type I or II portland cements (with moderate contents of

C_3A, C_3S) in combination with mineral admixtures should be suitable for making concrete mixtures for the marine environment. Alternatively, the use of blended portland cements, viz. Type IP containing up to 30% low-calcium fly ash or Type IS containing up to 70% granulated blast-furnace slag, ought to be considered especially because portland cements with high contents of C_3A, C_3S, and alkalis can be more safely used as blended cements. In summary, the author would like to reiterate that, with concrete sea structures any ordinary portland cement, typically containing 6–12% C_3A, should be satisfactory provided that the cement is compatible with the admixtures being used in the concrete mixture. When the use of mineral admixtures is not contemplated, Type IS and Type IP cements should be seriously considered from the standpoint of durability. ASTM Type III and V cements are generally not recommended owing to the high heat of hydration, and low rate of development of strength and watertightness, respectively.

Admixtures

One or more admixtures can be incorporated into a concrete mixture because the properties of both fresh and hardened concrete can be suitably modified by their use.[5–8] Some surfactants are suitable for air entrainment; others act as plasticizing agents. By plasticizing a concrete mixture, it is possible to reduce the water content at a given consistency; therefore, in the United States the plasticizing chemicals are called water-reducing agents. A major advancement in the 1970s was the development of high-range water-reducing agents, also called superplasticizers, which are able to produce cohesive concrete mixtures of flowing consistency with 20% to 30% less water content than the concrete without the superplasticizer. Evidently, superplasticizers are now playing a key role in the production of high-strength and low-permeability concretes. Soluble chemicals are frequently employed to control the time of set; for instance, calcium chloride is commonly used to accelerate the time of set, particularly under conditions of cold weather concreting.

The use of mineral additions either as admixtures to portland cement concrete mixtures or as ingredients of blended portland cements results in improvements in workability, impermeability, and resistance to thermal cracking and alkali–aggregate expansion.[4-7] Classification and characteristics of commonly used chemical and mineral admixtures are given in Chapter 3; so it is only a brief summary relevant to durability of concrete in the marine environment that is presented next.

Both ACI and FIP recommendations prohibit the use of calcium chloride or any admixtures containing chloride. There are numerous examples of chloride-related deterioration of reinforced and prestressed concrete structures due to corrosion of the embedded steel. For the same reason, many investigators including Hognestad[9] have warned against the use of *seawater as mixing water* in concrete. Reinforced concrete structures of a naval base in Bermuda started falling apart within 10 years of construction because seawater had been used as a mixing water. In a similar case at a petrochemical plant in the Gulf of Arabia, where seawater was used for mixing concrete, almost complete delamination of a reinforced concrete wall occurred in 10 years; on the other hand, at the same location concrete in ten *unreinforced* sea walls continues to be in good condition.

It is a common observation that microcracks in concrete, owing to various causes, tend to originate at the site of microstructural inhomogeneities, such as local areas of high water/cement ratio (e.g., the transition zone between cement paste and coarse aggregate or cement paste and steel) and poorly dispersed zones of cementitious particles. Therefore, concretes with low water content, especially those containing fine particles of a mineral admixture, require the use of a plasticizing or water-reducing chemical admixture for producing homogeneous and workable mixtures. Before the advent of superplasticizers, lignosulfonate-type plasticizers were mostly used for this purpose (see Table 1.1, Condeep Beryl A platform, 1975). A by-product of the wood pulp industry, most commercial lignosulfonate admixtures contain enough sugars and other impurities to cause excessive set retardation and air entrainment. High-purity superplasticizers of the naphthalene or melamine sulfonate-type are relatively free from such harmful side effects. Compared to lignosulfonates, naphthalene or melamine sulfonates are better dispersants for the cement–water system because the lignosulfonates exhibit a high degree of cross-linkage and form spherical microgel flocs, which are not as efficient in covering the surface of cement particles as the linear molecules of a superplasticizer (Fig. 6.1).

Lack of compatibility of a superplasticizing admixture with a particular brand of portland cement or with other admixtures present in the concrete mixture is a common phenomenon leading to problems, such as rapid slump loss and inadequate air entrainment. Whereas commercial products that are reportedly immune from these problems are now being marketed, it is desirable to undertake laboratory trials with locally available materials in order to resolve unforeseen problems before undertaking the actual construction. For instance, with the splashing zone concrete for Gullfaks C, Norwegian Contractors specified 3–5% air entrainment with 0·20 mm void spacing. In laboratory tests, the desired amount of air and void spacing could not be obtained by using the locally available naphthalene superplasticizer. The difficulty was resolved satisfactorily when a mixture of melamine and naphthalene sulfonate superplasticizers was used.

Air entrainment is now mandatory with concrete that is to be

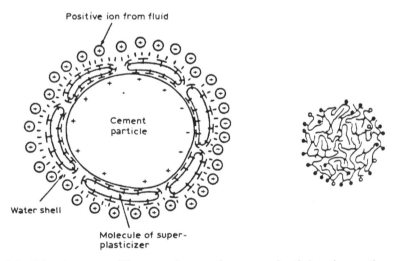

Fig. 6.1. Effectiveness of linear polymer of a superplasticizer in coating a cement particle (left), compared to a cross-linked lignosulfonate polymer (right).

subjected to repeated cycles of freezing and thawing. Surfactants derived from wood resins, proteinaceous products, and synthetic detergents can be effectively used to entrain closely spaced air voids in concrete. These air voids act as relief valves for the hydraulic pressure, which is generated during the frost action. To protect concrete from cracking due to freezing and thawing cycles, generally 5–7% entrained air voids with a minimum spacing factor of 0·2 mm (min. 24 mm^2/mm^3 surface area of air voids), is recommended. According to Moksnes,[10] adequate frost resitance was obtained for the splashing zone concrete for Gullfaks C platform with 3–5% air (spacing factor < 0·25 mm, and specific surface > 25 mm^2/mm^3).

Earlier, there were reports suggesting that very low water/cement ratio concrete mixtures containing condensed silica fume may have little or no freezable water to warrant protection from frost attack. However, a recent laboratory study by Pigeon et al.[11] has shown that even 0·3 water/cement ratio concrete mixtures with 9% condensed silica fume contained enough freezable water to cause damage in freezing and thawing tests unless air entrainment was used. With the materials and test conditions used by the authors, the value of critical spacing factor was found to be 0·4 mm for the silica fume concretes, and <0·4 mm for those without the silica fume.

Superplasticized concrete mixtures, especially those containing silica fume or other fine particles, normally require a high dosage of an air-entraining admixture for adequate void spacing. Also, fresh concrete mixtures tend to lose a part of the entrained air on placement and consolidation. As a safeguard against the loss of air, it is essential that the air content and the void spacing factor are periodically checked from core samples of hardened concrete.

In addition to frost resistance, it is observed that the entrainment of small air voids in a concrete mixture significantly improves the cohesiveness, workability, and homogeneity. On the other hand, since each 1% of entrained air lowers the compressive strength by approximately 5%, an overall 20–25% strength reduction due to air entrainment can be expected in ordinary concrete mixtures. In situations where high strength is as important as high impermeability, and where frost action is not of much concern, air entrainment is not a proper solution for obtaining highly workable and cohesive concrete mixtures. In such cases, it is more appropriate to use mineral admixtures.

The types and characteristics of commonly used mineral admixtures are covered by several recent publications.[7,12–17] Both pozzolanic and cementitious mineral admixtures are characterized by fine particle size (high specific surface), and either total absence of crystallinity (glassy structure) or poor crystallinity. Whereas the particle size plays an important role in determining the rheological characteristics of fresh concrete mixtures (cohesiveness, bleeding, workability, etc.), the degree of crystallinity influences the chemical activity. Low-calcium fly ashes, typically with 60–80% aluminosilicate glass and 300–500 m^2/kg surface area, behave like natural pozzolans; silica fume and rice husk ash comprised essentially of amorphous silica with 20 000–60 000 m^2/kg surface area, are not only superior pozzolans on account of their high reactivity, but owing to their high surface area are also very effective in controlling the bleeding tendency in freshly placed concrete mixtures. As discussed in Chapter 3, mixtures with excessive bleeding would produce a weak transition zone; microcracking in the transition zone increases the permeability and reduces the durability of concrete. High-calcium fly ashes and granulated blast-furnace slag typically contain 90% calcium aluminosilicate glass, and are cementitious and therefore can be used in larger proportions than pozzolanic admixtures.

By improving the workability and homogeneity of concrete, reducing the heat of hydration, and increasing the strength of the transition zone, the mineral admixtures are able to enhance the resistance of concrete to cracking from a variety of causes. Since both the strength and permeability of concrete in service are determined by the degree of internal microcracking, it should be obvious why the use of a suitable mineral admixture is so important when long-term durability of concrete in the hostile seawater environment is one of the primary design considerations.

The results from a recent experimental study by Moukwa,[18] on concrete specimens exposed to cold seawater in the tidal zone, have confirmed that the use of mineral admixtures reduced the deterioration of concrete by densifying the matrix and improving the transition zone (the paste–aggregate interfacial area). Air-entrained concrete mixture with 0·45 water to cementitious ratio were made, containing either of the two mineral admixtures, a condensed silica fume (8% by weight cement replacement with silica fume) or a blast-furnace slag (30% by weight cement replacement with slag). After proper curing the concrete specimens were exposed to severe thermal shock conditions by freezing in air at −25°C for 16 hours followed by thawing in seawater at −1°C for 8 hours. The change in the weight of the specimens and the amount of seawater absorbed after each cycle of thermal shock were used as the criteria for internal microcracking. The reference concrete containing no mineral admixtures showed the highest absorption of seawater and broke after the sixth thermal shock cycle. The concrete containing the silica fume showed no deterioration during the 15-cycle test. Not only did it show the least absorption, but also the amount of seawater absorbed decreased with increasing cycles of thermal shock. The author concluded that this effect was attributable to the dense matrix and strong transition zone which resisted microcracking. The concrete containing the blast-furnace slag absorbed more water than the silica fume concrete and broke after the ninth thermal shock cycle. The type and amount of slag used were probably inadequate to strengthen the transition zone since cracks occurred in most cases at the paste–aggregate interface.

Aggregates

Natural mineral aggregates, both coarse and fine, comprise over 90% of the total aggregate used for making concrete. Sand and gravel are generally produced by cleaning and grading naturally occurring deposits. When this is not economical, they are produced by rock crushing. Lightweight aggregates for making structural quality lightweight concrete are usually made by bloating shale, clay, or slate. In the United States, most codes of recommended practice require the normal-weight and the lightweight aggregates to meet ASTM C 33 and ASTM C 330 Standard Specifications, respectively. The specifications include grading requirements and limits on deleterious impurities. The list of deleterious substances includes clay lumps, silt, friable particles, coal and lignite, alkali-reactive minerals, iron sulfides, and decayed organic matter (see Chapter 3).

As discussed before, microcracking in the transition zone between coarse aggregate particles and cement paste is an important cause for the permeability of concrete in service. Since the aggregate size, shape, and mineralogy have a great influence on the strength of the transition zone, the effect of aggregate characteristics on concrete strength and durability should be obvious. Generally, aggregate particles larger than 20 mm, and either too elongated or too flat, tend to accumulate bleedwater near the surface by the wall effect. This results in a porous and weak transition zone, which microcracks easily under stress and thus becomes a contributing factor in lowering the strength and in increasing the permeability of concrete. Although ACI and FIP codes make no mention of the relationship between the maximum size of aggregate and strength or permeability of concrete (Table 6.1), the change in the North Sea concreting practice probably reflects this concern. Concrete mixtures for older structures contained up to 32 mm coarse aggregate particles; however, in recently built structures the maximum aggregate size has been limited to 20 mm.

In regard to the influence of aggregate type on concrete durability, theoretical considerations show that the aggregate which develops strong physical and chemical bonds with portland cement paste, and which has a similar elastic modulus and coefficient of thermal expansion as compared to the hydrated cement paste, should produce a more durable concrete. Judged by the foregoing criteria, sandstone and quartzitic gravels, when used as aggregate, would yield less durable concrete than crushed limestone or expanded clay or shale. From laboratory and field experience it has indeed been confirmed that both limestone and lightweight aggregates form a strong transition zone, which enhances the durability of concrete. Holm et al.[19] reported that compared to a normal-weight concrete there was less microcracking at the aggregate–cement paste interface in a lightweight aggregate concrete from normal heating–cooling and freezing–thawing cycles, to which the test concretes were exposed for more than 20 years.

Aïtcin and Mehta[20] recently reported that high-strength concrete specimens made either from a granite aggregate (which contained inclusions of a soft mineral) or a siliceous gravel aggregate gave lower strengths, lower values of the elastic modulus, and broader hysteresis loops in the elastic range than corresponding concrete specimens made either from a crushed limestone or from a basalt aggregate (Fig. 6.2). From the shape and size of hysteresis loops obtained by loading and unloading concrete specimens within the elastic range, the authors found that the aggregate mineralogy had a significant influence on the elastic behavior of concrete. It is suggested that microcracking in aggregate particles with weak inclusions or in the aggregate–cement paste transition zone (which is responsible for a broad hysteresis loop as a result of some inelastic deformation within the elastic loading range) would have an adverse effect on the long-term durability of concrete.

In short, from the standpoint of long-term durability of concrete in the seawater environment, the aggregates for making the concrete mixture should be mineralogically homogeneous, strong, hard, and clean. This is especially necessary for the splashing zone concrete, which is subject to heavy abrasion/erosion. From field experience with good quality concrete mixtures (30–40 MPa), it seems that under conditions of severe abrasion/erosion usually the coarse aggregate–cement paste bonding fails and the aggregates tend to pull out from the concrete matrix. Superplasticized, low water/cement ratio, concrete mixtures, containing highly active pozzolans (such as silica fume) have strong transition zones, and are therefore resistant to the aggregate pull-out phenomenon. When the aggregate particles are able to withstand the abrasive/erosive attack of the environment without being pulled out of the cement paste matrix, it is beneficial to use a high coarse/fine aggregate ratio in the concrete mixture (viz. 65/35) and an exceptionally hard rock as coarse aggregate. Gjørv et al.[21] reported that a very high-strength concrete mixture (150 MPa), containing silica fume and a superplasticizer, with jasper (quartz-diorite) as the coarse aggregate, showed an abrasion/erosion resistance of the same order as that of a granite test piece used for reference purposes.

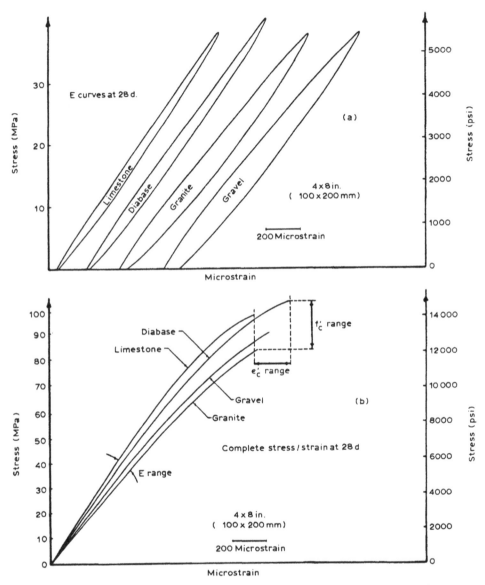

Fig. 6.2. Influence of aggregate type on the behavior of concrete specimens under uniaxial compression: (a) loading–unloading hysteresis loops in the elastic range, (b) stress–strain curves to failure. (Reproduced with permission from Ref 20.)

MIX PROPORTIONING OF CONCRETE MIXTURES

Mix proportioning is a process by which one arrives at the right combination of cement, aggregates, and admixtures for making a concrete mixture that would meet given requirements. With ordinary concrete mixtures the usual requirements are the consistency or slump of fresh concrete (one of the parameters governing the workability), and the 28-day uniaxial compressive strength, which is presumed to be an index of general concrete quality including the long-term durability. Of course, for special environmental conditions, such as exposure to freezing and thawing cycles, concrete has to be protected by air entrainment. Similarly, for corrosive environments, such as seawater, a maximum limit on the permeability of concrete may be specified. Since test methods for direct measurement of permeability are rather complex, this property may be controlled indirectly by specifying a maximum limit on the water/cement ratio.

With concrete-making materials of given characteristics and job specifications, such as slump and 28-d compressive strength, the variables generally under the control of a mix designer are the cement paste/aggregate ratio in the concrete mixture, the quality of the cement paste (i.e. water/cement ratio), the proportion between the fine and the coarse aggregate, and the type and dosage of admixtures to be used. As stated before, the task of mix proportioning is complicated by the fact that by changing a given variable some of the desired characteristics of concrete may be oppositely affected. For instance, an increase in water content of a stiff concrete mixture (with a given cement content) will no doubt improve the consistency of the fresh concrete, but at the same time reduce the strength of the hardened concrete. Nevertheless, the professional literature contains procedures derived from practical experience that are useful to arrive at optimum solutions in proportioning concrete mixtures. In the United States the most commonly used method for proportioning concrete mixtures is the ACI Committee 211 Recommended Practice. This eight-step method which involves the use of six tables, is described at the end of this chapter in Appendix A.

The principles underlying the ACI Recommended Practice can

be adapted for determining the proportions of concrete mixtures for most marine structures except those exposed to extremely hostile environments, as will be described later. The following illustration shows how this can be done.

1. *Choice of slump*: Concrete mixtures should have a consistency that permits thorough homogenization on mixing, and ease of transportation, placement, and consolidation without segregation. Note that superplasticized concrete for heavily reinforced sea structures to be placed by pumping generally requires 150–200 mm (6–8 in) slump.

2. *Choice of maximum size aggregate*: Owing to narrow spacing between bars in heavily reinforced structural elements and from the standpoint of keeping the permeability of concrete low, it would be desirable to limit the maximum aggregate size to 10 mm (3/8 in). Concrete mixtures with 25 to 37 mm maximum size aggregate can be used for unreinforced or lightly reinforced structures.

3. *Determination of mixing water content and air content*: With well-graded normal aggregates the amount of mixing water depends on the maximum aggregate size, the desired consistency and the air content of the concrete. Assuming that no air entrainment is needed, it can be determined from the ACI Table 2 in Appendix A that approximately 360 lb/yd³ (210 kg/m³) mixing water will be needed to produce 150–175 mm slump in a concrete mixture containing 19 mm maximum size aggregate. A 25% water reduction by the mandatory use of a superplasticizing admixture should bring down the water requirement to 270 lb/yd³ (160 kg/m³).

4. *Selection of water/cement ratio*: For structures exposed to seawater, the ACI Committee 201 (Table 4, Appendix A) recommends a maximum permissible water/cement ratio of 0·40, even when a higher water/cement ratio may be acceptable from consideration of the concrete strength. Suppose the specified 28-d concrete strength is 4500 psi (30 MPa). This corresponds to 5700 psi average strength (see Appendix A) which, according to Table 3 (Appendix A) would have called for 0·43 water to cement ratio.

5. *Calculation of cement content*: From the water content and the water/cement ratio in Steps 3 and 4 above, the calculated cement content is of the order of $270 \div 0 \cdot 4 = 675 \, \text{lb/yd}^3$ $(400 \, \text{kg/m}^3)$.[†]

6. *Selection of admixtures*: In addition to the high-range water-reducing admixture (superplasticizer), a high-quality pozzolan should be used to improve workability and reduce permeability. Approximately 5 to 15 liter/m³ (1 to 3 gal/yd³) of a naphthalene or melamine sulfonate type superplasticizer may be needed to obtain the desired consistency. Either condensed silica fume (7–10% by weight of cement) or a high quality fly ash (15% to 20% by weight of cement) should be considered for use as a pozzolanic *additive* to the concrete mixture.[†] From the standpoint of cost economy let us assume that fly ash will be used instead of silica fume. Note that 20% fly ash by weight of cement corresponds to $135 \, \text{lb/yd}^3$ $(81 \, \text{kg/m}^3)$.

7. *Estimation of coarse aggregate content*: The optimum amount of coarse aggregate in a concrete mixture depends on the aggregate grading and maximum size (see ACI Table 5, Appendix A). Assuming the fineness modulus of the available sand to be 3·0, the dry-rodded volume of fraction of coarse aggregate from Table 5 is 0·6. This corresponds to $1620 \, \text{lb/yd}^3$ $(970 \, \text{kg/m}^3)$ coarse aggregate when calculated in a manner similar to the one shown by the sample computations in Appendix A.

8. *Estimation of the fine aggregate content*: Since all other constituents of the mixture except the fine aggregate are known, either the absolute volume method or the unit weight method can be used to determine the fine aggregate content by difference. Using the unit weight method (ACI Table 6, Appendix A), with a unit weight of $3960 \, \text{lb/yd}^3$ $(2170 \, \text{kg/m}^3)$, the fine aggregate content needed is equal to $3960 - (270 + 675 + 135 + 1620) = 1260 \, \text{lb/yd}^3$ or $750 \, \text{kg/m}^3$.

[†] Under special circumstances, such as slow setting and hardening of concrete in cold weather, the cement content may be increased by 20% and the pozzolanic additive may be skipped. Some suitable steps may have to be taken to control thermal cracking due to heat of hydration in concrete mixtures with high cement content.

Steps 7 and 8 require a knowledge of the aggregate characteristics, such as the fineness modulus and bulk density. According to a simpler method for calculating the batch weight for the first laboratory trial, the unit weight of fresh concrete is assumed from previous experience. Knowing the water, cement, and pozzolan contents, this assumption makes it possible to calculate the content of total aggregate. Next, the coarse and fine aggregate proportions are determined on the assumption that 35% to 40% fine aggregate by weight of total aggregate would be sufficient for obtaining a concrete mixture with desired workability and high abrasion resistance. Such an assumption may not be valid for non-air-entrained concrete mixtures containing no pozzolanic admixture.

In Table 6.2 the ACI and FIP recommendations on concrete mix proportions are compared with the North Sea field practice. The minimum compressive strength requirements of both the ACI and FIP are far below the current North Sea concrete specifications, which are probably governed by the permeability rather than the compressive strength consideration. The 36-MPa concrete specified for the Ekofisk and Beryl A Condeep platforms, built in the early 1970s, barely qualifies as high-strength concrete. The specified strength of the Gullfaks C (1987–88) concrete was 56 MPa, which is approximately 50% higher than the specification for Beryl A concrete. The actual 28-d compressive strength of the Gullfaks C concrete cores was found to be approximately 70 MPa. It may be noted that, to some degree, there is a direct proportionality between strength and the coefficient of permeability.

Both the ACI and FIP recommendations call for a maximum 0·45 water/cement ratio concrete for the submerged zone and 0·40 for the splashing and atmospheric zones. The precast tunnel liners (Fig. 6.3) on the French side of the Channel Tunnel are made with a 55 MPa concrete mixture containing 400 kg/m³ cement content, 0·35 water/cement ratio (1·5% superplasticizer by weight of cement), and crushed limestone aggregate.[22] The average value of the coefficient of permeability of specimens cut from the cores is 4×10^{-13} kg/Pa ms. The concrete for the precast tunnel liners on the British side of the tunnel has a similar water/cement ratio but contains 310 kg/m³ ordinary portland cement and 130 kg/m³ fly ash.

Although 430 kg/m³ cement content was used for the Beryl A

Table 6.2. Concrete Mix Specifications for Offshore Structures.

Specification	ACI 357-84 Recommendation	FIP-1985 Recommendation	North Sea field practice
Compressive strength[a] minimum	35 MPa for all zones; 42 MPa where severe surface degradation is likely	32 MPa for all zones; 36 MPa where abrasion resistance is required	Specified strengths for Beryl A (1974) and Gullfaks C (1986–87) concretes were 36 MPa and 56 MPa, respectively. Actual strengths were about 20% higher than the specified value
Water/cement ratio maximum	0·45 for the submerged zone, and 0·40 for the splash zone and the atmospheric zone	0·45 maximum, but 0·40 is preferred	0·41 for Beryl A concrete. 0·38 for Gullfaks C concrete
Cement content minimum	356 kg/m^3 (600 lb/yd^3)	320 kg/m^3 and 360 kg/m^3 for 40 mm and 20 mm max. aggregate size, respectively. 400 kg/m^3 for the splash zone	430 kg/m^3 cement was used for Beryl A concrete, and 380 kg/m^3 for Statfjord C (1981)
Consistency	No requirement	No requirement	Minimum 120 mm slump was specified for Beryl A concrete. A much higher slump (240 mm) is specified for Gullfaks C owing to a 25% increase in the steel reinforcement
Permeability minimum	No requirement	No requirement	10^{-13} to 10^{-14} kg/Pa ms

[a] This is 28-d, average compressive strength from 150 mm dia. by 300 mm long standard-cured test cylinders. It should be noted that in the range of 30 to 60 MPa concretes, the cylinder strength, typically, is about 80% of the 150 mm cube strength.

concrete, such high cement contents are no longer considered desirable from the standpoint of thermal cracking. The concrete mixtures for Statfjord C and Gullfaks C structures contained 380 kg/m^3 cement content. The splashing zone concrete for Gullfaks C also contained 5% silica fume by weight of cement, with enough superplasticizer to reduce the water/cementitious ratio to 0·38. More importantly, the trend for directly specifying a minimum permeability coefficient in offshore concrete structures is worthy of notice. Although the ACI and FIP recommendations

(a)

(b)

Fig. 6.3. Precast concrete tunnel liners for the Channel Tunnel: (a) French side; (b) British side. (Courtesy: P. Poitevin)

have no such requirement as yet, the concrete mixtures for the North Sea platforms are now required to have a permeability coefficient of the order of 10^{-13} to 10^{-14} kg/Pa ms (Table 6.2).

CONCRETE MIXTURES FOR EXTREMELY HOSTILE ENVIRONMENTS

Coastal and offshore marine structures in the splashing zone are subjected to tremendous impact from storm waves and floating objects (such as ice floes in the Arctic). From the standpoint of long service life concrete mixtures with exceptionally high resistance to abrasion/erosion, impact, and fatigue are required for use under such extremely hostile conditions. High-performance concrete mixtures with abrasion resistance similar to granite panels and with extremely low permeability have recently become available for commercial use. Typically, these mixtures contain sufficient dosage of a superplasticizer to give high workability even at 0·30, or less, water to cement ratios. The uniaxial compressive strength is usually in the 60 to 120 MPa range, although such high strengths may be neither called for nor utilized in the structural design. The microstructure of the hardened concrete is dense owing to the absence of large capillary voids and a relatively strong transition zone which has a much lower tendency for microcracking, when compared to ordinary concrete. Therefore, plain high-strength concrete is brittle in the sense that, unlike ordinary concrete, its energy absorption capacity through the enlargement of microcracks is low. Usually, to obtain adequate toughness or ductility, the high-strength concrete structures are heavily reinforced and prestressed.

High-performance concrete mixtures cannot be proportioned with the help of currently available methods, including the procedure recommended by the ACI Committee 211. With a given combination of materials, usually extensive laboratory trials are needed before suitable mixture proportions are arrived at. To reduce the number of laboratory trials for determining the mixture proportions for a high-performance concrete, several methods have been proposed in the recently published literature. Since there is considerable interest from the marine concrete industry in these mixtures, a method developed by Mehta and Aïtcin[23] is described below.

The authors define high-performance concrete as a material which is not only characterized by high strength and exceptionally low permeability (e.g., 1×10^{-14} kg/Pa ms), but also has high dimensional stability. To achieve a high dimensional stability it is necessary to reduce the magnitude of drying shrinkage and creep strains by limiting the total content of cement paste in concrete and by the use of a coarse aggregate which has high elastic modulus. From theoretical considerations, the authors concluded that 35% cement paste by volume represents the optimum cement paste content in balancing the conflicting requirements of low permeability and high dimensional stability.

Although compressive strength directly is not an important criterion for high-performance concrete, the authors noted that concretes with very low permeability and high dimensional stability normally possess greater than 60 MPa strength and, therefore, the compressive strength of concrete can be used as a basis of proportioning, and quality control of concrete mixtures. Using high-quality natural aggregates, it is possible to modify the composition and properties of the cement paste to produce concretes of up to 120 MPa compressive strength. For the purpose of mix-proportioning the authors recommend that 60 to 120 MPa strength range be divided into several strength grades. Since a general inverse relationship exists between strength and the water content of the concrete mixture, the authors have assumed that this relationship can be exploited for the proportioning of concrete mixtures.

In regard to the type and dosage of mineral admixtures, Mehta and Aïtcin[23] suggest one of the three following options. According to the first, portland cement alone may be used without any mineral admixtures. This option should only be used when necessary. Under most conditions the second option is recommended, which involves a partial cement replacement with a high quality fly ash or ground granulated iron blast-furnace slag. For a variety of reasons, with concrete mixtures of 100 MPa or higher strength, mineral admixtures must be used. For calculating the first trial batch, the authors suggest a 75:25 volumetric proportion between portland cement and the selected mineral admixture. At ordinary temperature the improvements in strength and impermeability resulting from the use of fly ash or slag do not manifest themselves until at least 28 days of moist curing. This process can be accelerated by substituting 10% of the fly ash or slag with a more reactive pozzolan (viz. condensed silica fume or rice husk ash), which is the third option.

The amount of reduction in the water content and the improvement in workability (typically 150 to 250 mm slump without segregation) that is often required of high-performance concrete mixtures can only be obtained by the use of superplasticizing admixtures. Most superplasticizers are sulfonated derivatives of either naphthalene or melamine, and are commercially available in the form of solutions, the normal dosage being 0·8% to 2% solid by weight of the cementitious materials. The water present in the superplasticizer solution must be accounted for when calculations are made for the mixing water.

To calculate the contents of coarse and fine aggregates the authors take advantage of the fact that the total aggregate volume is known (65% of the concrete mixture) and that, with concrete mixtures containing a high cement content and mineral admixtures, generally a 2:3 ratio between the fine aggregate and the coarse aggregate is sufficient from the standpoint of workability. Based on the assumptions discussed above, Mehta and Aïtcin have recommended the following eight-step procedure for computing the weight of component materials needed for the first trial batch for a 1 m^3 non-air-entrained concrete mixture.

Step-by-Step Procedure

Step 1: Choice of Strength
The 60–120 MPa strength range is arbitrarily divided into five strength grades, namely 65, 75, 90, 105, and 120 MPa, average strength of standard-cured concrete specimens at 28-d. In general, the weather-resisting characteristics of concrete will improve with

increasing strength, but so will the cost. Assuming that the local aggregate is not strength-limiting, the choice of strength should be made from a cost–benefit analysis.

Step 2: Estimation of Mixing Water

For a given strength grade, Table 6.3 is used to estimate the maximum content of mixing water. This estimate is based on experience with high-slump superplasticized concrete mixtures, containing 12–19 mm maximum size aggregate (MSA). By correcting for the extra water present in the superplasticizer (also in aggregates, if wet), the batch mixing water can be calculated (see Steps 6 and 7 below).

Step 3: Volume Fraction of Cement Paste Components

Since the total volume of cement paste is $0 \cdot 35 \text{ m}^3$, subtracting the mixing water content (Step 2) and $0 \cdot 02 \text{ m}^3$ entrapped air content, the calculated volumes of the total cementitious material for each Strength Grade are shown in Table 6.4. Also shown in Table 6.4 are the volume fractions of portland cement and mineral admixtures, assuming one of the following three options:

Option #1:   Portland cement (PC) alone
Option #2:   Portland cement plus either fly ash (FA) or blast-furnace slag (BFS) in 75:25 ratio by volume.
Option #3:   Portland cement + either FA or BFS + CSF (condensed silica fume) in 75:15:10 ratio by volume, respectively.

Table 6.3. Relationship between Average Compressive Strength and Maximum Water Content.

Strength grade	Average strength (MPa)	Maximum water content (kg/m^3)
A	65	160
B	75	150
C	90	140
D	105	130
E	120	120

Table 6.4. Volume Fraction of Components in 0·35 m³ Cement Paste (m³).

Strength grade	Water	Air	Total cementitious materials	Option #1 PC only	Option #2 PC + (FA or BFS)	Option #3 PC + (FA or BFS) + CSF
A	0·16	0·02	0·17	0·17	0·127 5 + 0·042 5	0·127 5 + 0·025 5 + 0·017 0
B	0·15	0·02	0·18	0·18	0·135 0 + 0·045 0	0·135 0 + 0·027 0 + 0·018 0
C	0·14	0·02	0·19	0·19	0·142 5 + 0·047 5	0·142 5 + 0·028 5 + 0·019 0
D	0·13	0·02	0·20	[a]	0·150 0 + 0·050 0	0·150 0 + 0·030 0 + 0·020 0
E	0·12	0·02	0·21	[a]	0·157 5 + 0·052 5	0·157 5 + 0·031 5 + 0·021 0

[a] Concrete grades D and E are not made without mineral admixtures.

Step 4: Estimation of Aggregate Content

From the total aggregate volume (0·65 m³), assuming 2:3 volumetric ratio between the fine and the coarse aggregate for Grade A mixture, the individual volume fractions will be 0·26 m³ and 0·39 m³, respectively. For other grades (B–E), due to the decreasing water content and increasing superplasticizer content, somewhat lower fine/coarse aggregate volume ratios may be assumed with increasing strength, for instance, 1·95:3·05 for Grade B, 1·90:3·10 for Grade C, 1·85:3·15 for Grade D, and 1·80:3·20 for Grade E.

Step 5: Calculation of Batch Weights

Typical specific gravity values for normal portland cement, fly ash (Class C) or blast-furnace slag, and condensed silica fume are 3·14, 2·5, and 2·1. Typical specific gravity values for natural siliceous sand and most normal-weight gravels or crushed rocks are assumed to be 2·65 and 2·70, respectively. Using the data from Step 3 (Table 6.4) and Step 4, the calculated SSD (saturated–surface dry) weights are shown in Table 6.5. For significant differences in the aggregate specific gravity compared to the assumed values, appropriate corrections should be made.

Step 6: Superplasticizer Dosage

If there is no prior experience with the superplasticizer, it is suggested that you start with 1·0% superplasticizer (on anhydrous solid basis) by weight of cementitious materials. When the specific gravity of the superplasticizer solution and the weight fraction of

Table 6.5. Mix proportions (SSD) for the First Trial Batch (kg/m^3)

Strength grade	Average strength (MPa)	Option	Cementitious materials			Total^a water	Coarse agg.	Fine agg.	Total batch	W/C
			PC	FA or BFS	CSF					
A	65	1	534	—	—	160	1 050	690	2 434	0·30
		2	400	106	—	160	1 050	690	2 406	0·32
		3	400	64	36	160	1 050	690	2 400	0·32
B	75	1	565	—	—	150	1 070	670	2 455	0·27
		2	423	113	—	150	1 070	670	2 426	0·28
		3	423	68	38	150	1 070	670	2 419	0·28
C	90	1	597	—	—	140	1 090	650	2 477	0·23
		2	447	119	—	140	1 090	650	2 446	0·25
		3	447	71	40	140	1 090	650	2 438	0·25
D	105	—	—	—	—	—	—	—	—	—
		2	471	125	—	130	1 110	630	2 466	0·22
		3	471	75	42	130	1 110	630	2 458	0·22
E	120	—	—	—	—	—	—	—	—	—
		2	495	131	—	120	1 120	620	2 486	0·19
		3	495	79	44	120	1 120	620	2 478	0·19

^{a} Total water includes the water in the superplasticizing admixture, the dosage of which may range from 10 to 20 liters/m^3, depending on consistency and strength requirements.

solids in the solution are known, it is easy to estimate the volume of the solution for a given batch. The following example shows how the calculations are made.

For Strength Grade A (65 MPa) with Option #3, the total weight of cementitious materials will be 500 kg/m³. For the trial batch, 1% superplasticizer solids, therefore, amounts to 5 kg/m³. If the weight fraction of solids in the solution is 40%, the weight of the solution is 5 divided by $0.4 = 12.5$ kg/m³. If the specific gravity of the solution is 1.2, the volume of the solution is 12.5 divided by $1.2 = 10.4$ liters/m³. The amount of water in the solution must be subtracted from the mixing water (Step 2), as discussed next.

Note that the weight of water in 10.4 liters/m³ superplasticizer solution is equal to $10.4 \times 1.2 \times 0.6 = 7.5$ kg/m³.

Step 7: Moisture Correction

Since the mix proportions in Table 6.5 are on S.S.D. basis, depending on the moisture condition of batch aggregates the appropriate moisture corrections in both fine and coarse aggregate must be made. A corresponding correction in the batch mixing water in Table 6.5 is also made for the amount of water present in the superplasticizer solution (see Step 6).

Step 8: Trial Batch Adjustment

Because of the many assumptions underlying the proposed method, the calculated mix proportions for the first trial batch serve only as a guide. Several laboratory trials using the actual materials may have to be made before one arrives at the right combination of materials and mix proportions which satisfy the given criteria of workability and strength. The following suggestions are offered for the purpose of making the needed adjustments.

If the concrete mixture is too stiff, the superplasticizer dosage should be gradually increased until the desired consistency is obtained. Increase in the superplasticizer dosage may be accompanied by certain unwanted effects, such as a tendency for segregation and/or set retardation. The former may be corrected by increasing the ratio between the fine and coarse aggregate or by using a fine aggregate with a lower fineness modulus. For mixtures with no mineral admixtures, this problem may be resolved by incorporation of silica fume or fly ash or both. To solve the set retardation problem, combinations of other types or brands of locally available superplasticizers and cements should be tested. Modified superplasticizers containing set accelerators, which claim not to cause excessive set retardation, are also available commercially. It should be noted that when the cause of stiff consistency of a concrete mixture is a high content of reactive C_3A in the cement, additional dosages of superplasticizers alone may not work. In such cases, an increase in the water/cement ratio becomes unavoidable.

If the 28-d compressive strength of the trial mix is lower than the specified average strength, an examination of the fracture surface of concrete specimens and stress–strain curves may provide a clue to the weakest component of the composite. In the case of frequent debonding between cement paste and coarse aggregate particles, either the cement paste in the transition zone needs strengthening or the aggregate is too smooth. In the former case, control of bleeding by reduction in the water content and/or incorporation of fine particles of a suitable mineral admixture may solve the problem. In the latter case, the use of crushed aggregate should be considered.

REFERENCES

1. Verbeck, G. J., Field and laboratory studies of the sulfate resistance of concrete, *Proceedings Thorvaldson Symposium on Performance of Concrete,* University of Toronto Press, 1971, pp. 113–24.
2. Mehta, P. K., Durability of concrete in marine environment—a review, *Performance of Concrete in Marine Environment,* ed. V. M. Malhotra, ACI SP-65, 1980, pp. 1–20.
3. Rosenburg & Gaidis, Letters to the Editor, *Concrete International,* **12**(5) (1990), 12.
4. Mehta, P. K., Durability of concrete exposed to marine environment—a fresh look, *Performance of Concrete in Marine Environment,* ed. V. M. Malhotra, ACI SP-109, 1988, pp. 1–30.

5. *Concrete Admixtures Handbook,* ed. V. S. Ramachandran, Noyes Publications, Park Ridge, N.J., 1984.

6. *Concrete Admixtures: Use and Applications,* ed. M.. R. Rixom, The Construction Press, London, 1978.

7. *Supplementary Cementing Materials,* ed. V. M. Malhotra, CANMET, Ottawa, Canada, 1988.

8. Mehta, P. K., *Concrete: Structure, Properties, and Materials,* Prentice Hall, Englewood Cliffs, N.J., 1986, pp. 249–81.

9. Hognestad, E., *Proceedings Gerwick Symposium on Durability of Concrete in Marine Environment,* Dept. of Civil Engineering, University of California at Berkeley, 1989, p. 126.

10. Moksnes, J., Oil and gas concrete platforms in the North Sea—reflections of two decades of experience, *Proceedings Gerwick Symposium on Durability of Concrete in Marine Environment,* Dept. of Civil Engineering, University of California at Berkeley, 1989, pp. 127–46.

11. Pigeon, M., Gagne, R. & Foy, C., Critical air-void spacing factor for low water/cement ratio concretes with and without condensed silica fume, *Cement and Concrete Research,* **17** (1987) 896–906.

12. Swamy, R. N. (ed.), *Cement Replacement Materials,* Surrey University Press, Guildford, 1986, 259 pages.

13. Helmuth, R., *Fly Ash in Cement and Concrete,* Portland Cement Association, 1987, 203 pages.

14. Rilem Final Report Committee 73-SBC, *Materials and Structures J.,* **21** (121), Chapman and Hall, London (1988), 69–80.

15. ACI Committee 226, Silica fume in concrete, *ACI Materials J.,* March–April (1987), 158–66.

16. ACI Committee 226, Ground granulated blast-furnace slag, *ACI Materials J.,* July–August (1987), 327–42.

17. ACI Committee 226, Fly ash in concrete, *ACI Materials J.,* September–October (1987), 381–408.

18. Mouwka, M., Deterioration of concrete in cold sea waters, *Cement and Concrete Research,* **20** (1990), 439–46.

19. Holm, T. A., Bremner, T. W. & Newman, J. W., Lightweight aggregate concrete subject to severe weathering, *Concrete International,* **6** (1984), 49–54.

20. Aïtcin, P. C. & Mehta, P. K., Effect of coarse aggregate characteristics on mechanical properties of high-strength concrete, *ACI Materials J.,* March–April (1990), 103–7.

21. Gjørv, O. E., Baerland, T. & Ronning, H. R., High strength concrete for highway pavements and bridge decks, *Proceedings Conference on Utilization of High Strength Concrete, Stavanger, Norway,* 1987, pp. 111–22.

22. Poitevin, P., Concrete for long service life: Channel Tunnel, *Dansk Beton*, No. 2 (1990), 71–6.

23. Mehta, P. K. & Aïtcin, P. C., Principles underlying the production of high performance concrete, *ASTM J. Cement, Concrete and Aggregates*, Winter (1990), 70–8.

APPENDIX A: ACI COMMITTEE 211 RECOMMENDED PRACTICE FOR COMPUTING PROPORTIONS OF CONCRETE MIXTURES

Step 1: Choice of slump
If the slump is not specified, a value appropriate for the work can be selected from Table 1. Mixes of the stiffest consistency that can be placed and compacted without segregation should be used.

Step 2: Choice of maximum size of aggregate
For the same volume of the coarse aggregate, using a large maximum size of a well-graded aggregate will produce less void space than using a smaller size, and this will have the effect of reducing the mortar requirement in a unit volume of concrete. Generally, the maximum size of coarse aggregate should be the largest that is economically available and consistent with the

Table 1. Recommended Slump for Various Types of Construction.

Types of construction	Slump (in)	
	Maximum[a]	Minimum
Reinforced foundation walls and footings	3	1
Plain footings, caissons, and substructure walls	3	1
Beams and reinforced walls	4	1
Building columns	4	1
Pavements and slabs	3	1
Mass concrete	2	1

[a] *May be increased 1 in for methods of consolidation other than vibration. Reproduced with permission from the American Concrete Institute.*

dimensions of the structure. In no event should the maximum size exceed one-fifth of the narrowest dimension between the sides of the forms, one-third the depth of slabs, or three-fourths of the minimum clear spacing between reinforcing bars.

Step 3: Estimation of mixing water and the air content
The quantity of water per unit volume of concrete required to produce a given slump is dependent on the maximum particle size, shape, and grading of the aggregates, as well as on the amount of entrained air; it is not greatly affected by the cement content of the concrete mixture. If data based on experience with the given aggregates are not available, assuming normally-shaped and well-graded particles, an estimate of the mixing water with or without air entrainment can be obtained from Table 2 for the purpose of computing the trial batches. The data in the table also show the approximate amount of entrapped air expected in

Table 2. Approximate Mixing Water and Air Content Requirements for Different Slumps and Maximum Sizes of Aggregates.

Slump (in)	Water (lb/yd^3 of concrete for indicated maximum sizes of aggregate)						
	$\frac{3}{8}$ in	$\frac{1}{2}$ in	$\frac{3}{4}$ in	1 in	$1\frac{1}{2}$ in	2 in	3 in
Non-air-entrained concrete							
1–2	350	335	315	300	275	260	240
3–4	385	365	340	325	300	285	265
6–7	410	385	360	340	315	300	285
Approximate amount of entrapped air in non-air-entrained concrete (%)	3	2·5	2	1·5	1	0·5	0·3
Air-entrained concrete							
1-2	305	295	280	270	250	240	225
3–4	340	325	305	295	275	265	250
6–7	365	345	325	310	290	280	270
Recommended average total air content (%)	8	7	6	5	4·5	4	3·5

Reproduced with permission from the American Concrete Institute.

non-air-entrained concrete and recommend levels of air content for concrete in which air is to be purposely entrained for frost resistance.

Step 4: Selection of water/cement ratio
Since different aggregates and cements generally produce different strengths at the same water/cement ratio, it is highly desirable to develop the relationship between strength and water/cement ratio for the materials actually to be used. In the absence of such data, approximate and relatively conservative values for concretes made with Type I portland cement can be taken as shown in Table 3. Since the selected water/cement ratio must satisfy both the strength and the durability criteria, the value obtained from the table may have to be reduced depending on the special exposure requirements (Table 4).

Step 5: Calculation of cement content
The required cement content is equal to the mixing water content (Step 3) divided by the water/cement ratio (Step 4).

Table 3. Relationships between Water–Cement Ratio and Compressive Strength of Concrete.

Compressive strength at 28 days (psi)[a]	Water–cement ratio, by weight	
	Non-air-entrained concrete	Air-entrained concrete
6 000	0·41	—
5 000	0·48	0·40
4 000	0·57	0·48
3 000	0·68	0·59
2 000	0·82	0·74

[a] *Values are estimated average strengths for concrete containing not more than the percentage of air shown in Table 2. For a constant water–cement ratio, the strength is reduced as the air content is increased. Strength is based on 6 by 12-in cylinders moist-cured 28 days at 73·4 ± 3°F (23 ± 1·7°C) in accordance with Section 9(b) of ASTM C 31, for Making and Curing Concrete Compression and Flexure Test Specimens in the Field.*
Reproduced with permission from the American Concrete Institute.

Table 4 Maximum Permissible Water–Cement ratio for Concrete in Severe Exposures.

Type of structure	Structure wet continuously or frequently and exposed to freezing and thawing[a]	Structure exposed to seawater or sulfates
Thin sections (railings, curbs, sills, ledges, ornamental work) and sections with less than 1 in cover over steel	0·45	0·40[b]
All other structures	0·50	0·45[b]

[a] Concrete should also be air-entrained.
[b] If sulfate-resisting cement (Type II or Type V of ASTM C 150) is used, the permissible water cement ratio may be increased by 0·05.
Based on report of ACI Committee 201, Durability of Concrete in Service. Reproduced with permission from the American Concrete Institute.

Step 6: Estimation of coarse-aggregate content

Economy can be gained by using the maximum possible volume of coarse aggregate on a dry-rodded basis per unit volume of concrete. Data from a large number of tests have shown that for properly graded materials, the finer the sand and the larger the size of the particles in coarse aggregate, the more is the volume of coarse aggregate that can be used to produce a concrete mixture of satisfactory workability. It can be seen from the data in Table 5 that, for a suitable degree of workability, the volume of coarse aggregate in a unit volume of concrete is dependent only on its maximum size and the fineness modulus of the fine aggregate. It is assumed that differences in the amount of mortar required for workability with different aggregates, due to differences in particle shape and grading, are compensated for automatically by differences in dry-rodded void content.

The volume of aggregate, in cubic feet, on a dry-rodded basis, for 1 yd^3 of concrete is equal to the volume fraction obtained from Table 5 multiplied by 27. This volume is converted to the dry weight of coarse aggregate by multiplying by its dry-rodded unit weight.

Table 5. Volume of Coarse Aggregate per Unit of Volume of Concrete.

Maximum size of aggregate (in)	Volume of dry-rodded coarse aggregate[a] per unit volume of concrete for different fineness moduli of sand			
	2·40	2·60	2·80	3·00
$\frac{3}{8}$	0·50	0·48	0·46	0·44
$\frac{1}{2}$	0·59	0·57	0·55	0·53
$\frac{3}{4}$	0·66	0·64	0·62	0·60
1	0·71	0·69	0·67	0·65
$1\frac{1}{2}$	0·75	0·73	0·71	0·69
2	0·78	0·76	0·74	0·72
3	0·82	0·80	0·78	0·76
6	0·87	0·85	0·83	0·81

[a] *Volumes are based on aggregates in dry-rodded condition as described in ASTM C 29, Unit Weight of Aggregate. These volumes are selected from empirical relationships to produce concrete with a degree of workability suitable for usual reinforced construction. For less workable concrete such as required for concrete pavement construction they may be increased about 10%. For more workable concrete, such as may sometimes be required when placement is to be by pumping, they may be reduced up to 10%.*
Reproduced with permission from the American Concrete Institute.

Step 7: Estimation of fine-aggregate content

At the completion of Step 6, all the ingredients of the concrete have been estimated except the fine aggregate; its quantity is determined by difference, and at this stage either the 'weight' method or the 'absolute volume' method can be followed.

According to the *weight method,* if the unit weight of fresh concrete is known from previous experience, then the required weight of fine aggregate is simply the difference between the unit weight of concrete and the total weights of water, cement, and coarse aggregate. In the absence of a reliable estimate of the unit weight of concrete, the first estimate for a concrete of medium richness (550 lb of cement per cubic yard, medium slump of 3 to 4 in) and approximately 2·7 aggregate specific gravity can be obtained from Table 6. Experience shows that even a rough estimate of the unit weight is adequate for the purpose of making trial batches.

Table 6. First Estimate of Weight of Fresh Concrete.

Maximum size of aggregate (in)	First estimate of concrete weight[a] (lb/yd³)	
	Non-air-entrained concrete	Air-entrained concrete
$\frac{3}{8}$	3 840	3 690
$\frac{1}{2}$	3 890	3 760
$\frac{3}{4}$	3 960	3 840
1	4 010	3 900
$1\frac{1}{2}$	4 070	3 960
2	4 120	4 000
3	4 160	4 040
6	4 230	4 120

[a] *Values calculated for concrete of medium richness (550 lb of cement per cubic yard) and medium slump with aggregate specific gravity of 2·7. Water requirements based on values for 3 to 4 in of slump in Table 2. If desired, the estimated weight may be refined as follows when necessary information is available: for each 10-lb difference in mixing water from the Table 2 values for 3 to 4 in of slump, correct the weight per cubic yard 15 lb in the opposite direction; for each 100-lb difference in cement content from 550 lb, correct the weight per cubic yard 15 lb in the same direction; for each 0·1 by which aggregate specific gravity deviates from 2·7, correct the concrete weight 100 lb in the same direction.*
Reproduced with permission from the American Concrete Institute.

In the case of the *absolute volume method* the total volume displaced by the known ingredients (i.e. water, air, cement, and coarse aggregate) is subtracted from the unit volume of concrete to obtain the required volume of fine aggregate. This in turn is converted to weight units by multiplying it by the density of the material.

Step 8: Adjustments for aggregate moisture

Generally, the stock aggregates are moist; without moisture correction the actual water/cement ratio of the trial mix will be higher than selected by Step 4, and the saturated–surface dry (SSD) weights of aggregates will be lower than estimated by Steps 6 and 7. The mix proportions determined by Steps 1 to 7 are therefore assumed to be on an SSD basis. For the trial batch, depending on the amount of free moisture in the aggregates, the mixing water is reduced and the amounts of aggregates correspondingly increased, as will be shown by the sample computations.

Step 9: Trial batch adjustments

Due to so many assumptions underlying the foregoing theoretical calculations, the mix proportions for the actual materials to be used must be checked and adjusted by means of laboratory trials consisting of small batches (e.g., 0·01 yd³ of concrete). Fresh concrete should be tested for slump, workability (freedom from segregation), unit weight, and air content; specimens of hardened concrete cured under standard conditions should be tested for strength at the specified age. After several trials, when a mixture satisfying the desired criteria of workability and strength is obtained, the mix proportions of the laboratory-size trial batch are scaled up for producing full-size field batches.

Sample Computations

Job Specifications

Type of construction	Reinforced concrete footing
Exposure	Mild (below ground, not exposed to freezing or sulfate water)
Maximum size of aggregate	$1\frac{1}{2}$ in
Slump	3 to 4 in
Specified 28-day compressive strength	3 500 psi

Characteristics of the Materials Selected

	Cement Lone Star, Type I	Fine aggregate, Felton, No. 2	Coarse aggregate, Fair Oaks, Gravel
Bulk specific gravity	3·15	2·60	2·70
Bulk density (lb/ft³)	196	162	168
Dry-rodded unit weight (lb/ft³)	—	—	100
Fineness modulus	—	2·8	—
Moisture deviation from SSD condition (%)	—	+2·5	+0·5

Steps 1 to 7: Computing mix proportions (SSD basis, lb/yd³)

Step 1. Slump = 3 to 4 in (given).

Step 2. Maximum aggregate size = $1\frac{1}{2}$ in (given).

Step 3. Mixing water content (non–air–entrained concrete) = 300 lb. Approximate amount of entrapped air = 1% (Table 2).

Step 4. Average strength from equations in the following section, 'Methods of determining average compressive strength from the specified strength (assuming 300 psi standard deviation from past experience) = $3500 + 1·34 \times 300 = 3900$ psi. Water/cement ratio (Table 3) = 0·58.

Step 5. Cement content = $300/0·58 = 517$ lb.

Step 6. Volume fraction of gravel on dry-rodded basis (Table 5) = 0·71

> Dry-rodded volume of gravel = $0·71 \times 27 = 19·17$ ft³
> Weight of gravel = $19·17 \times 100 = 1917$ lb.

Step 7. Using the *weight method*: unit weight of concrete (Table 6) = 4070 lb/yd³

Weight of sand = $4070 - (300 + 517 = 1917) = 1336$ lb.

Using the *absolute volume method*:

Volume displaced by water = 300/62·4	=	4·81 ft³
Volume displaced by cement = 517/196	=	2·64 ft³
Volume displaced by gravel = 1917/168	=	11·43 ft³
Volume displaced by air = $27 \times 0·01$	=	0·27 ft³
total		19·15
Volume displaced by sand = $(27 - 19·15)$	=	7·85 ft³
Weight of sand = $7·85 \times 162 = 1272$ lb.		

Since the absolute volume method is more exact, the proportions determined by this method will be used.

Step 8: Moisture Adjustment for the Laboratory Trial Batch

Material	SSD (lb/yd³)	SSD (lb/0·01 yd³)	Moisture correction (b)	Mix proportions for the first trial batch (lb)
Cement	517	5·17		5·17
Sand	1 272	12·72	$12·72 \times 0·025 = 0·3$	13·02
Gravel	1 917	19·17	$19·17 \times 0·005 = 0·1$	19·27
Water	300	3·00	$3 - (0·3 + 0·1)$	2·60
Total	4 006	40·06	←—must be equal—→	40·06

Step 9: *Making the First Laboratory Trial and Adjusting the Proportions*
Measured properties of fresh concrete from the first trial batch:

$$\text{slump} = 4\tfrac{3}{4}\,\text{in}$$
$$\text{workability} = \text{slight tendency to segregate and bleed}$$
$$\text{unit weight} = 148\,\text{lb/ft}^3\ (3996\,\text{lb/yd}^3)$$
$$\text{air content} = 1\%$$

Action taken for the second trial batch: reduce the gravel by $\tfrac{1}{4}$ lb
and increase the sand by the same amount.
Batch weights for the second trial batch:

$$
\begin{aligned}
\text{cement} &= \ 5\cdot17\,\text{lb} \\
\text{sand} &= 13\cdot27\,\text{lb} \\
\text{gravel} &= 19\cdot02\,\text{lb} \\
\text{water} &= \ 2\cdot60\,\text{lb} \\
\hline
&\ \ 40\cdot06\,\text{lb}
\end{aligned}
$$

Measured properties of fresh concrete from the second trial batch:

$$\text{slump} = 4\,\text{in}$$
$$\text{workability} = \text{satisfactory}$$
$$\text{unit weight} = 148\,\text{lb/ft}^3$$
$$\text{air content} = 1\%$$

Three 3- by 6-in cylinders were cast and moist cured at
$73\cdot4 \pm 3°\text{F}$.
Average 28-day compressive strength was 4250 psi, with less than
5% variation in strength between the individual cylinders.
Recalculated mix proportions for the full-size field batch are as
follows.

	Present stock (lb/yd³)	Moisture correction (for conversion to SSD condition) (lb)	SSD basis (lb/yd³)
Cement	517		517
Sand	1 327	1327 × 0·025 = 33	1 294
Gravel	1 902	1902 × 0·005 = 10	1 892
Water	260	260 + (33 + 10)	303
Total	4 006	←——must be equal——→	4 006

Methods of Determining Average Compressive Strength from the Specified Strength[†]

ACI 322, *Building Code Requirements for Structural Plain Concrete*, and ACI 318, *Building Code Requirements for Reinforced Concrete*, specify that concrete shall be proportioned to provide an average compression strength (f'_{cr}) which is higher than the specified strength (f'_c) so as to minimize the probability of occurrence of strengths below f'_c.

When a concrete production facility has a suitable record of 30 consecutive tests of similar materials and conditions expected, the standard deviation can be calculated in accordance with the expression

$$S = \left[\frac{\Sigma (x_i - \bar{x})^2}{n - 1} \right]^{1/2}$$

where S is the standard deviation (psi), x_i the strength value from an individual test, $\bar{x}$ the average strength of n tests, and n the number of consecutive strength tests. When data for 15 to 25 tests are available, the calculated value of the standard deviation may be modified according to the following data:

Number of tests	Multiplication factor
15	1·16
20	1·08
25	1·03

The required average compressive strength (f'_{cr}), which is to be used as the basis for calculating concrete mix proportions, shall be the larger of equation (1) or (2):

$$f'_{cr} = f'_c + 1·34S \qquad (1)$$

$$f'_{cr} = f'_c + 2·33S - 500 \qquad (2)$$

Equation (1) provides a probability of 1 in 100 that averages of three consecutive tests will be below the specified strength f'_c. Equation (2) provides a similar probability of individual tests being more than 500 psi below the specified strength.

[†] Based on ACI Building Code 318.

When adequate data are not available to establish a standard deviation, the required average strength can be determined from the following:

Specified compressive strength, f'_c (psi)	Required average compressive strength, f'_{cr} (psi)
Less than 3 000	$f'_c + 1\,000$
3 000 to 5 000	$f'_c + 1\,200$
Over 5 000	$f'_c + 1\,400$

Chapter 7
Concreting Practice

For making durable concrete structures the selection of proper materials and mix proportions is only the first step. Sufficient attention must also be paid to the concrete production and construction practice. Reflecting the growing awareness in the concrete construction industry that the *neonatal history of concrete* (early age from birth) plays an important part in determining the service life, the author[1] has stated:

> In the medical profession it is well recognized that in order to develop into a healthy person a newborn baby needs special attention during the early period of growth. Something similar applies to concrete, although there is no clear definition of how early is the early age. Concrete technologists agree that deficiencies acquired by fresh concrete due to the loss of workability at or before placement, segregation and bleeding during consolidation, and an unusually slow rate of hardening (strength gain) can impair a concrete permanently and reduce its service life.

In the case of concrete, usually the 'early age' period is limited to the first 24 hours after production. A detailed review of the recommended concrete production and construction practices is beyond the scope of this book. However, from the standpoint of concrete durability, a summary of some of the relevant and helpful information is presented below.

BATCHING, MIXING, AND TRANSPORT

Water and liquid admixtures can be batched either by volume or by weight; however, the solid components of a concrete mixture are more accurately batched by weight. In fact, most concrete today is batched and mixed in ready-mixed concrete plants equipped with automatic or semi-automatic batching controls. Truck-mixers rather than centralized mixers are used predominantly in the United States, because in truck mixing all three operations—batching, mixing, and transporting of concrete to the job site—can be combined into one operation which can be handled by the same equipment. However, central mixing is being increasingly preferred from the standpoint of homogeneity and uniformity of the product. Inclined-axis central mixers of the revolving drum type, either with rear or front discharge, are commonly used in the ready-mixed concrete industry. High-speed mixers are also being investigated because of the significant improvement in the cohesiveness of freshly mixed concrete and the strength of hardened concrete, made possible by the thoroughness of the mixing operation.

The *sequence* of mixing, i.e. the order in which the components of a concrete mixture are batched, can have a considerable influence on the properties of the resulting product. Inadequate deflocculation of particles of cement as well as mineral admixtures, such as silica fume, has been observed in slow-speed mixers when cement and silica fume are added to dry-batched aggregates before the addition of water and superplasticizer. Well-homogenized mixtures showing better strength characteristics are obtained when cement and silica fume or other pozzolanic additions are first turned into a completely deflocculated slurry by high-speed agitation with mixing water containing the superplasticizer. The aggregates are introduced subsequently to this slurry. Also, since high-slump concrete tends to undergo relatively greater slump loss during transport, it is recommended to withhold a part of the mixing water and approximately one-third of the intended dosage of the superplasticizer, which can be added at the job site just before the concrete placement.

Adequate air-entrainment is frequently a problem with high-consistency concrete mixtures, especially those containing fine particles, such as silica fume. Moksnes *et al.*[2] resolved this problem by a modified batching procedure, which required the addition of the air-entraining agent at a somewhat stiffer consistency of concrete than was finally desired. Since the presence of a melamine-type superplasticizer had a positive influence on the stability of the air void system, a mixture of melamine and naphthalene sulfonates and a special mixing procedure had to be used to assure satisfactory air entrainment for the splashing zone concrete of the Gullfaks C offshore platform.

Transportation of freshly mixed concrete to the job site should be carried out as quickly as possible to minimize too much loss of consistency which could hamper subsequent operations, namely the placement, consolidation, and finishing of concrete. Under mild or cold weather conditions, there is usually a negligible loss of consistency during the first 30 minutes after the addition of mixing water to cement. Concrete kept in a slow state of agitation, such as in a ready-mixed concrete truck, would undergo a small slump loss with time; but normally this does not present any serious problem provided the placement and consolidation are completed within $1\frac{1}{2}$ hours of mixing.

Depending on the type and range of work, various kinds of equipment are available for transportation of concrete to the job site.[1] Trucks, belt conveyors, bucket cranes, chutes, and elevators are among the equipment commonly used by the industry for concrete transporting. In choosing the equipment, besides cost the primary objective is to ensure that the concrete mixture will not segregate during the transport. For instance, with ready-mixed concrete trucks, a continuous agitation of the entire concrete batch must be assured. This means that the blades of the truck mixer must be regularly checked for excessive wear so that there is no significant build-up of segregated concrete inside the mixer.[3]

PLACEMENT, CONSOLIDATION, AND FINISHING

At the job site, to prevent segregation the concrete should be placed as near as possible to its final resting position. Generally, fresh concrete is deposited in horizontal layers of uniform thickness and each layer is thoroughly consolidated by vibrators before the next layer is placed. The rate of placement is carefully controlled so that the layer immediately below is still in the plastic state when a new layer is deposited. This prevents the formation of cold joints or planes of potential delamination, which occur when fresh concrete is placed on a concrete that has already hardened.

Placement by pumping is becoming increasingly popular in the concrete industry. When excessive pumping pressures are needed, a part of the mixing water may be lost by surface adsorption to aggregate particles. It could lead to 'slump loss', and consequently other problems, such as poor consolidation and finishing. To resolve this difficulty at the job site, stiff concrete mixtures are retempered with additional water. Provision for this additional moisture in concrete should be made in mix proportioning, otherwise durability and strength will be adversely affected owing to the higher water/cement ratio. Also, invariably there is some loss of entrained air during placement by pumping. Again, the solution is to provide for this loss in mix-proportioning specifications, and to regularly inspect the cores from hardened concrete for proper air content and void spacing.

Consolidation is the process of molding concrete in forms and around embedded parts, and elimination of air bubbles entrapped during the mixing operation. Immersion-type vibrators are commonly employed for this purpose, as they are quite effective in molding even relatively stiff concrete mixtures. Consolidation of concrete by vibration is also helpful in obtaining a uniform distribution of solids and water in the concrete mixture. Removal of trapped air voids and water-pockets from the aggregate surface is beneficial to concrete strength and durability.

When flowing or self-leveling concrete mixtures containing superplasticizers were first introduced, it was speculated that no consolidation would be necessary for such high-consistency concrete mixtures. However, experience shows that high-slump (200–250 mm) superplasticized mixtures tend to be thixotropic and, unless vibrated, they will not properly fill narrow spaces in heavily reinforced concrete elements. Forssblad[4] reported that, without consolidation, reinforced concrete showed a tendency to entrap 3% to 5% air, a reduction in bond strength, and settlement cracks around reinforcing steel. Consolidation by vibration is therefore essential even for superplasticized concrete mixtures. However, with superplasticized concrete the radius of action for an immersion vibrator is 20% to 30% greater than with ordinary concrete of plastic consistency and, therefore, the vibration time can be reduced.

Inhomogeneous distribution of solids, liquid, and air in concrete is undesirable, and may be caused both as a result of over-vibration or under-consolidation. Excessive segregation due to too much over-vibration should be avoided. However, Whiting et al.[5] found that over-consolidation to the point of incipient segregation resulted in higher compressive strength and lower chloride permeability. Also, it should be noted that sometimes *revibration* of concrete, approximately an hour after the initial consolidation but before setting, is helpful in welding successive castings together and in removing microcracks or voids created by settlement and bleeding, especially around the reinforcing steel. A report by ACI Committee 309 contains detailed information on aspects of concrete consolidation, such as imperfections, under-vibration, over-vibration, and revibration.[6]

Generally, it is flatwork such as slabs and pavements which require proper *finishing* to produce a dense surface that will remain maintenance-free for a long time. However, in the case of massive piers and beams also it is highly desirable to obtain a smooth finish, free from surface voids and defects. This would have the effect of reducing the permeability of the surface layer, called *skincrete*, which serves as the first line of defense against the ingress of corrosive fluids. Sadegzadeh et al.[7] reported that the abrasion resistance of a concrete was greatly influenced by the pore structure of the cement paste matrix in the outermost 2 to 3 mm surface zone of concrete. The pore structure was affected by the water/cement ratio and the surface finishing procedures. Repeated trowelling, after the moisture brought out to the surface by initial power trowelling was allowed to evaporate, gave the most dense pore structure and best abrasion resistance. Concrete mixtures with high cement contents, and especially those containing condensed silica fume, tend to be sticky and hard to finish. In such cases, special vibratory screeds should be used to obtain a smooth and impermeable surface.

It may be noted that precipitation of aragonite [$CaCO_3$], and brucite [$Mg(OH)_2$], on the concrete surface as a result of seawater exposure also produces an impermeable skin. Buenfeld and Newman[8] reported that the range of pressures encountered by offshore structures have a negligible effect on this skin, although temperature variations and the presence of organic compounds have a significant effect.

CURING AND FORMWORK REMOVAL

Proper curing is very important to obtain the desired strength, impermeability, and long-term durability in concrete. The purposes of concrete curing are, first to prevent the loss of moisture which is needed for cement hydration and, second to control the temperature of concrete for a period long enough to achieve a desired maturity (strength) level. Above freezing temperatures, flatwork can be cured by ponding, whereas columns, beams, and other structures are usually cured by fogging or by covering with wet burlap. These methods provide some cooling effect through evaporation, which is beneficial in the case of massive structures, especially in hot weather. With thermally cured products special precautions may have to be taken to provide humid conditions for preventing the moisture loss. In France, the precast concrete segments for lining the Channel Tunnel were exposed to warm air (up to 35°C) and the concrete temperature was permitted to rise to a maximum of 60°C within 5 to 6 hours after casting. Since the segments were not covered, steam was injected in order to maintain the high humidity required for proper curing.

When the ambient temperatures are below freezing, concrete must be protected with insulating blankets and cured with steam or electrically heated forms or infrared lamps. Alternatively, in cold weather, the temperature of fresh concrete may be raised by heating the mixing water and/or aggregates. FIP recommends that concrete temperature at the time of placement should be at least 5°C and should be maintained above this point until the concrete has reached a minimum of 5 MPa compressive strength.

Another approach for concrete curing, which is increasingly used by the construction industry, is based on prevention of moisture loss by sealing the surface of concrete immediately after the finishing operation through the application of waterproof curing paper, polyethylene sheets, or membrane-forming coatings. Whenever there is doubt that the concrete surface will not remain moist through the recommended curing period, or when there is a danger of thermal cracking due to the use of very cold curing water, it is desirable to consider this approach.

Asselanis *et al.*[9] have reported that high-strength, low-permeability, concrete mixtures of the type currently being used in sea structures, become virtually impermeable within 3 to 7 days of moist curing; thereafter the continuation of moist curing has no effect on concrete properties. According to Gerwick,[3] because most of the concrete mixtures used for sophisticated offshore structures are highly impermeable, the emphasis today is primarily on sealing the concrete surface against moisture loss rather than on moist curing. However, the author warns that with membrane curing compounds the heat from the sun or from cement hydration may degrade the curing compound and, therefore, one or more additional applications may be necessary during the first day. Also, the white pigmented varieties are especially suitable for reflecting heat in hot climates.

Whether curing is achieved by membrane sealing or by water spray, it must be emphasized again that the maintenance of proper humidity conditions for a certain length of time after placement of a concrete mixture is as essential for obtaining the desired impermeability in hardened concrete as the selection of proper materials and mix proportions. As stated earlier, from the standpoint of long-term durability, it is absolutely essential to begin with a totally impermeable concrete surface (skincrete), which cannot be achieved without proper finishing and adequate curing. It may be noted that tests for surface hardness are more reliable for evaluation of the impermeability of skincrete than strength tests.

Seawater should not be used for curing of reinforced or

prestressed concrete elements. According to the current Norwegian practice, it is permitted to expose high-strength concrete members to seawater after three days of proper curing, because by this time the concrete is assumed to have become essentially impermeable.

Curing and formwork removal are the last operation during concrete construction which can adversely influence strength, permeability, and other important properties of the product regardless of the quality of materials for concrete-making and the mix proportions used. The time of formwork removal has considerable economic significance because early removal of forms permits a faster turn-around and keeps the construction cost low; on the other hand, concrete structures are known to have failed when forms were removed before the concrete had gained sufficient strength. As a rule of thumb, forms should not be removed until concrete is strong enough to carry both the dead load and any imposed load during construction. Also, since skincrete is so important as a first line of defense against corrosive fluids, concrete should be sufficiently hard to prevent any damage to the surface during the form removal. This is especially important in formwork removal during which thermal cracks are likely to occur on exposure of a relatively warm concrete to an icy breeze or a very cold curing water. Under such conditions, it is desirable to insulate the concrete surface immediately after the formwork removal.

Evaluation of *in situ* strength of concrete in a structure should not be based on laboratory-cured specimens unless temperature and moisture conditions in the laboratory curing are similar to the field curing conditions. Maturity meters and other nondestructive techniques are now commonly available for direct assessment of concrete strength before formwork removal, and are more reliable than techniques based on measurement of surface hardness. Under normal curing conditions (23°C and 100% relative humidity), ordinary concrete mixtures made with ASTM Type I portland cement may gain in 1-day adequate strength for form removal (approximately 5 MPa compressive strength). Concrete mixtures with high cement contents or those containing a rapid-hardening cement or an accelerating admixture may require less time.

CRACK WIDTHS AND COVER THICKNESS

Sufficient concrete cover over the steel reinforcement and prestressing tendons is important for durability, particularly in the marine environment. To find an optimum value for the cover thickness it should be remembered that too much cover increases the width of cracks in the cover concrete, whereas too little will lead to easier penetration of the corrosive salt water. It is assumed that owing to normal volume changes, reinforced concrete will crack in service; however, the designers tend to limit the crack widths in the belief that there is a relationship between the crack width and corrosion. Although it is known now that a direct relationship does *not* exist between structural cracks and corrosion, wide cracks tend to enhance the rate of corrosion in permeable concrete. As cited by Burdall and Sharp,[10] Lenshow found increasing probability for serious corrosion when crack widths approached 0·5 mm and little corrosion in the range of 0·1 to 0·3 mm, which is recommended by most industry codes. ACI 224 R-80 recommends 0·15 mm maximum crack width at the tensile face of reinforced concrete structures subjected to a wetting and drying environment or seawater spray. FIP recommends that crack widths at points nearest to the main reinforcement should not exceed 0·2 mm for 50 mm cover, or 0·3 mm for 75 mm cover. Also, it is generally observed that from the standpoint of corrosion of steel the cracks longitudinal to the reinforcement are more critical than transverse cracks.

For offshore and coastal structures, with 0·5 m or more thickness, the nominal cover thickness for protection of reinforcing steel and prestressing tendons is related to the exposure zone of the structure. For the submerged zone both FIP and ACI 357 R recommend a 50 mm cover thickness over principal reinforcement, and 75 mm over prestressing tendons. In the splashing and the atmospheric zones, which are subject to seawater spray, the recommended cover thickness is 65 mm for reinforcing steel and 90 mm for prestressing tendons. Stirrups may have 13 mm less cover than the preceding recommended values.

As stated earlier, when concrete cracks, the crack width is found to be directly proportional to the cover thickness. Since large crack widths are undesirable because microcracking in service due to a combination of causes (see Chapter 4) will increase the permeability of concrete, structural designers are beginning to feel concerned about the large cover depths and the corresponding large crack widths. According to Gerwick (personal communication), with the type of modern highly impermeable concrete mixtures used for offshore platforms, it should not be necessary to have more than 50 mm cover thickness for either reinforcing steel or prestressing steel. Burdall and Sharp[10] compared the nominal cover thickness recommended by various industry codes (Table 7.1). It should be noted that the proposed guideline draft of the UK Department of Energy is in conformity with Gerwick's views on the subject of cover thickness.

Joints

Case histories of deteriorated concrete in old structures show that many problems originate from water seepage through construction joints and cold joints. To obtain watertight construction joints, the surface should be carefully prepared by sand blasting or a high-pressure water jet until all dirt and laitance are removed and the coarse aggregate is exposed to a depth of about 6 mm. Cold joints between successive concrete placements can be pre-

Table 7.1. Comparison of Cover Requirements.

	Nominal cover to reinforcement (mm)								
	DEn 3rd Edn	DnV	Lloyds	FIP 3rd edn	FIP 4th edn	BS 8110	BS 6235	ACI	UK Dept. of Energy, Draft Guidance Note
Reinforcing bars									
(a) Submerged	60	50	60	60	50	60	60	50	45
(b) Splash	75	50	75	75	75	60	75	65	70
(c) Atmospheric	75	40	75	75	75	60	75	50	55
Prestressing tendons									
(a) Submerged	75	100	75	75	65	60	75	75	45
(b) Splash	100	100	100	100	90	60	100	90	70
(c) Atmospheric	100	80	100	100	90	60	100	75	55

Reproduced, with permission, from Ref. 10.

vented by the use of set-retarding admixtures in concrete, and proper vibration.

Thermal Cracks

Thermal stresses in concrete due to the heat of hydration of cement must be reduced by preventing high temperature gradients, which usually arise 24 to 48 hours after the concrete placement. Tensile stresses are developed when the element begins to cool but is restrained from shrinking. Since the tensile strength of concrete is low, too high a thermal gradient will lead to high tensile stresses and eventually cause cracking of the element. Thermal cracking is probably the most important factor contributing to an increase in the permeability and a reduction in general durability of otherwise good-quality concrete mixtures, unless adequate measures have been taken to protect the concrete from steep thermal stress gradients. According to FIP Recommendations, when the minimum dimension of concrete to be placed at one time is greater than 600 mm, and the cement content of the mixtures is more than $400 \, kg/m^3$, precautions for the control of thermal cracking should be taken.

Reducing the temperature rise in concrete is a major problem, particularly with mixtures containing high cement contents and with elements of the rather massive size that is required for many sea structures. Heat of hydration may be reduced by limiting the cement content, and by controlling the cement composition and fineness. The cement content can be reduced by lowering the water/cement ratio as much as possible through control of aggregate grading, and type as well as dosage of the water-reducing admixture, while maintaining the desired workability, strength, and permeability of concrete. Portland–slag cement and portland–pozzolan cement generally give a lower heat of hydration than normal portland cement, though portland–slag cements containing very finely ground particles of a reactive blast-furnace slag may give a high heat of hydration. Coarse-ground portland cements, such as ASTM Type II cement, normally produce a lower heat of hydration compared to the general-purpose, ASTM Type I, portland cement. The adiabatic temperature rise in mass

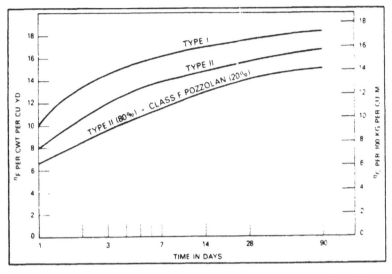

Fig. 7.1. Rate of adiabatic temperature rise in mass concrete (Ref. 11).

concrete mixtures made with different types of portland cements is shown in Fig. 7.1.[11] A further reduction in the adiabatic temperature rise can be achieved by the use of blended portland–pozzolan (Type IP) or portland–slag cements. Instead of using blended portland cement, a lowering of the heat of hydration can also be achieved by the use of pozzolanic and cementitious admixtures in concrete.

Pre-cooling the fresh concrete mixture is now a widely used practice to prevent steep temperature gradients in mass concrete. The temperature of the concrete mix is usually reduced by one or more of the following methods:

1. Mixing shaved ice instead of water.
2. Introduction of liquid nitrogen into the concrete mixture.
3. Shielding aggregate piles from the sun; and water spray to cause cooling of aggregate by evaporation.

In the Norwegian offshore concrete practice, for large and insulated pours, the typical temperature rise can be 12 to 14°C per 100 kg cement. Thus a concrete mixture containing 400 kg/m³ cement content, will undergo a temperature change of the order of 50 to 60°C. For the splashing zone concrete for Gullfaks C, although partial reduction of cement content was achieved by the use of silica fume, crushed ice was also used to cool the concrete mixture. A further reduction in the cement content may be possible by a partial cement substitution with a high-quality fly ash.

In regard to thermal cracking, a sound construction practice that is frequently overlooked is the protection of young precast concrete members from too early an exposure to cold air. Gerwick[12] reports that after two years of service in the Arctic weather in Beaufort Sea, two of the heavily reinforced precast concrete caissons (600 to 1000 mm thick) showed extensive cracking on the outer face. The author suspects that some or all of these cracks originated during manufacture when young concrete was exposed to cold winds and these cracks were later amplified by the effect of wetting–drying and freezing–thawing cycles on water held in the very small cracks.

Thermal cracking is also known to occur owing to cycles of temperature extremes, such as in the Middle East where daily changes of up to 30°C are not uncommon. Gerwick[12] reported vertical cracks due to thermal strains and pole driving stresses in the prestressed concrete cylinder piles of the 26-km long Ju' Aymah LPG Trestle, which was built in 1978 using high-quality concrete. Unfortunately, there was inadequate spiral reinforcement to close the cracks once they occurred. The use of adequate reinforcement under unusual conditions of temperature or loading (viz. fatigue from berthing and moving loads from vessels) is a part of good concreting practice because microcracking of concrete resulting from any cause would reduce the durability of the material and the service life of the structure.

REFERENCES

1. Mehta, P. K., *Concrete: Structure, Properties, and Materials*, Prentice Hall, Englewood Cliffs, NJ, 1986, pp. 301–42.
2. Moksnes, J., Huag, A. K., Moder, M. & Bergvam, T., Concrete quality in Norwegian offshore structures, *Proceedings, Conference on High-Strength Concrete, Stavanger, Norway*, 1987, pp. 405–16.

3. Gerwick, B. C., *Design of Offshore Concrete Structures*, Wiley–Interscience, New York, 1986.

4. Forssblad, L., Need for consolidation of superplasticized concrete mixes, *Consolidation of Concrete*, ed. S. H. Gebler, ACI SP-96, 1987, pp. 19–38.

5. Whiting, D., Seegebrecht, W. G. & Tayabji, S., Effect of degree of consolidation on some important properties of concrete, *ACI SP-96*, 1987, pp. 125–60.

6. ACI Committee 309 Report, *ACI J. of Materials*, **84**(5) (1987), 410–49.

7. Sadegzadeh, M., Page, C. L. & Kettle, R. J., Surface microstructure and abrasion resistance of concrete, *Cement and Concrete Research*, **17** (1987), 581–90.

8. Buenfeld, N. R. & Newman, J. B., The development and stability of surface layers on concrete exposed to seawater, *Cement and Concrete Research*, **16** (1986), 721–32.

9. Asselanis, J., Aïtcin, P. C. & Mehta, P. K., Effect of curing conditions on the compressive strength and elastic modulus of very high-strength concrete, *ASTM J. Cement, Concrete, and Aggregates*, **11**(1) (1989), 80–3.

10. Burdall, A. C., and Sharp, J. V., Some aspects of revisions to the U.K. guidance note for offshore concrete structures, *Proceedings, International Conference on Concrete in the Marine Environment*, The Concrete Society, London, 1986, pp. 37–48.

11. ACI Committee 207 Report, *Concrete International*, **2**(5) (1980), 49.

12. Gerwick, B. C., International experience in the performance of marine concrete, *Concrete International*, **12**(5) (1990), 47–53.

Chapter 8
Repair of Marine Structures

Concrete structures are generally designed on the assumption that they will require no maintenance during their service life. The experience with reinforced concrete structures in the marine environment shows that this assumption is not valid. Almost everywhere in the world it is possible to find coastal structures, even some relatively new ones, which contain cracked or spalled concrete usually involving corrosion of the reinforcement.

As pointed out earlier, the corrosion of steel by seawater is generally neither the first nor the only cause of concrete cracking. Case histories of some of the deteriorated concrete structures described below will show that the corrosion of reinforcing steel usually follows other causes of concrete cracking. A network of pre-existing cracks in concrete provides the necessary conditions for acceleration of the corrosion–cracking–corrosion cycle, which eventually leads to serious deterioration of concrete. Rusty (orange or brown) stains along longitudinal cracks in concrete, which are usually parallel to the reinforcing steel, provide the visible evidence for corroding steel. The corrosion and cracking are generally in an advanced state when rust marks are discovered. This is why the corrosion of steel in concrete is mistakenly held to be the primary or even the sole cause of concrete cracking.

Some of the 600 mm hollow core octagonal concrete piles of the Rodney Container Terminal at the Saint John Port in New Brunswick, Canada, showed signs of distress soon after completion of the facility in 1975. Although there was evidence of

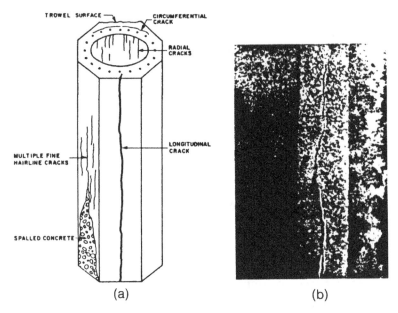

Fig. 8.1. (a) Diagrammatic representation of main types of pile distress at Rodney Terminal; (b) photograph of longitudinal cracks (reproduced, with permission, from Ref. 1).

corrosion of the reinforcing steel, the vertical cracks in the precast concrete piles (Fig. 8.1) were attributed mainly to high thermal gradients during the steam curing operation at the time of manufacture and to thermal stresses from cycles of freezing and thawing during the winter months, since the concrete mixture did not contain proper air-entrainment.[1] In 1978, 29 of the 1750 piles were repaired by epoxy grouting, and eleven severely cracked piles were repaired using fiber-glass jackets. In 1982, seven piles had to be replaced, and in 1983–84, 90 piles were repaired.[2]

In 1980, eleven of the 22 spandrel beams, 7·9 by 3·7 by 1·8 m, of the 17-year-old San Mateo–Hayward Bridge in Northern California had to be repaired at a heavy cost because of serious cracking of concrete that was attributed to corrosion of the reinforcing steel. These beams were made of precast concrete. It was noticed that the other eleven beams, which did not suffer any deterioration under similar conditions, were made of cast-in-place concrete. Since a high-quality concrete mixture (370 kg/m^3 cement content,

0·45 water/cement ratio) had been used in both the cases, it was suspected that a combination of heavy reinforcement and differential cooling (following steam curing during the manufacture) must have produced microcracks in the precast beams. Corrosion of the reinforcing steel began when these microcracks became continuous owing to the severe weathering exposure in the splashing zone.[3]

The Al-Shindagha tunnel under the Dubai Estuary in the Arabian Gulf, built in 1973–75, suffered leakage due to shrinkage cracks in the outside walls and leaking joints.[4] Subsequently, owing to corrosion of the reinforcement by seawater, which caused more cracking and enormous spalling of concrete, repairs had to be undertaken in 1986.

In 1979, about 15 years after the completion of the Zealand Bridge across the Eastern Scheldt Sea in southern Netherlands, deterioration of concrete cover in the bridge girders was noticed.[4] In various places corrosion of the reinforcing steel was discovered from rust stripes and from spalling of the cover. Investigation showed that the concrete quality was satisfactory; however, owing to defective workmanship the concrete cover on the reinforcement was 10 to 20 mm thick instead of the 30 mm required by the design specifications. This is a typical example showing that costly repairs to concrete sea structures can be avoided by inspection for 'preventive maintenance', rather than waiting for 'corrective maintenance', after serious deterioration has taken place.

For a variety of reasons, notable advancements in repair and rehabilitation of concrete structures have occurred during the last two decades. The main impetus to repair technology has come from numerous cases of deterioration in old concrete structures, such as bridge piers, bridge decks, tunnels, and navigation locks. Owing to the rising cost of new construction, in most instances it is more economical to extend by 15 to 20 years the service life of an older structure by repairing it rather than by complete replacement. Recent advancements in repair materials and methods have resulted in sophisticated technologies for rehabilitation of concrete structures that have either been accidentally damaged or deteriorated by natural weathering action. The repair of vertical faces of a concrete sea structure in the submerged zone (i.e. underwater repairing) and the tidal zone present special challenges, which have been successfully met by the use of anti-washout admixtures or by shotcreting with superplasticized concrete containing silica fume, as will be discussed.

The following operations, which are normally undertaken in a major repair, will be discussed in this chapter:

- evaluation of the condition of concrete before repair.
- determination of the causes of damage or deterioration.
- selection of the repair materials.
- selection of repair methods, including the method for surface preparation.

EVALUATION OF CONCRETE BEFORE REPAIR

Evaluation of the condition of concrete before repair is necessary in order to assess the extent of damage or deterioration. This information is useful for estimating the repair cost which, in turn, helps in making the decision on the three possible courses of action, namely replacement, repair, or no action. If repair happens to be the most cost-effective solution, a decision has also to be made regarding the urgency of repair.

Evaluation of concrete may also yield valuable information about the causes of damage. The damage due to sudden impact or overloading is easy to assess; however, the damage due to weathering or other causes of slow deterioration requires the use of special investigation techniques, which are described below. Information on causes of damage is one of the inputs needed for determining the purpose of repair, such as impermeability to a corrosive fluid or resistance to abrasion. For proper selection of repair materials and methods, it is essential to start with a clear understanding of repair objectives.

It is also important to make a note of the expected environmental conditions during the repair work. For selection of the right repair materials for a given job, it is necessary to know whether the surface of concrete to be repaired will be dry or wet, and whether the concrete temperature will be cold or warm. The ease or difficulty of access to the job will influence the method to be selected for execution of the repair. The techniques used for evaluation of the condition of concrete before repair are described next.

Background Information
It is useful to gather as much background information as possible from the review of original drawings and material specifications, and any history of previous repairs.

Visual Inspection
It is a good practice to begin with visual observations on visible signs of distress, such as cracks, spalls, rust spots, and leakage marks. Visual inspection alone is not sufficient for determining the need and methodology of repair. It is a preliminary screening process which helps to determine the type of nondestructive and laboratory tests to be undertaken for obtaining further information that will be needed before any decisions can be made. It is advisable to supplement the findings of visual inspection with photographs, video tapes, and sketches. For underwater survey, traditional visual inspection by a diver or an underwater camera may be adequate for many jobs; however, these methods may not be satisfactory for low-visibility conditions.

Robotics and Remote Sensors
According to Gerwick,[5] new survey techniques using robotics and higher resolution sounding systems offer a range of options and versatility, which were not possible before. A remote-operated vehicle (ROV) originally developed in the Netherlands for inspection work on the Eastern Scheldt Barrier Project, is shown in Fig. 8.2. It is a 6 m long by 4 m wide bottom-crawling vessel, designed to conduct underwater surveys in very turbid water with velocities up to 2·5 m/s. This vessel has six television cameras fitted with special lenses, four high-frequency acoustic transducers, and a sonar. It is designed to work at water depths ranging from 15 to 45 m.

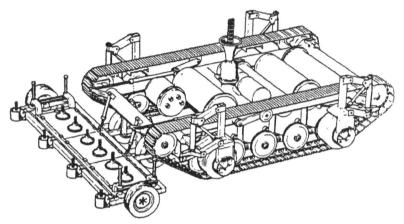

Fig. 8.2. The underwater survey vehicle 'Portunus' (reproduced, with permission, from Ref. 5).

Where large vehicles are inappropriate, a no-bottom-contact type vehicle may be employed. For the Eastern Scheldt Barrier Project, for inspecting the gap between the foundations of the precast piers and the sand layer an underwater self-propelled and free floating ROV was developed. It is tubular in shape (900 mm long by 42·5 mm diameter), and is equipped with lights, cameras, and pressure sensors.[5]

Soundings
Soundings are taken by striking the concrete surface with an ordinary hammer to locate areas of loose material, such as delamination of the concrete cover due to freezing and thawing cycles and/or corrosion of the reinforcement. This method, though easy and simple, can furnish valuable qualitative information on the condition of concrete to an experienced inspector; however, it is of limited value for underwater evaluation because of the diminished hearing ability in the presence of noise from waves and currents.

Surface Hardness
Surface hardness measurement methods provide quick, inexpensive, and nondestructive means of checking the uniformity

of quality of in-place hardened concrete. ASTM C 805 describes a procedure involving a rebound hammer, which gives the surface hardness reading by impacting a spring-loaded hammer against the concrete surface. ASTM C 803 describes a procedure involving a powder-activated driver that is used to fire a hardened-alloy probe into concrete. The size of indentation provides a measure of the surface hardness. However, the surface hardness measurement devices have limited value for underwater inspection of concrete.

Ultrasonic Testing

Ultrasonic testing is used to detect cracks, flaws, or voids. A low-frequency wave is sent from a transducer through a known distance of the material to a receiver, and the travel time of the wave is recorded. The presence of voids and cracks manifests itself as delay of velocity of travel time of the ultrasonic wave.

ASTM Standard Method C 597 is frequently used for ultrasonic pulse velocity testing, and commercially available instruments have been modified for underwater use. Although the data are quantitative, they should be treated with caution when evaluating the general uniformity of concrete, which is also influenced by the aggregate content and the location of reinforcing steel.

According to Gerwick,[5]

> 'Ultrasonics can be very useful to determine the degree of deterioration of the surface of the abraded concrete. Although this surface is frequently worn away to expose a clean sound surface, this may not be the case where reinforcing steel bars are exposed, where inserts and joints occur, or where impact or cavitation has caused cracking, spalling, and pitting. Ultrasonic inspection techniques can be difficult to apply to concrete due to the inhomogeneity resulting from the aggregates, which act as discontinuities when interacting with sound waves. Therefore, to inspect the surface of concrete for cracks and local pitting, it is helpful to use ultrasonic waves that do not penetrate too deeply into the concrete. Two such techniques are acoustic microscopy, and the leaky Rayleigh wave method.

Acoustic Microscopy

In acoustic microscopy, which is somewhat similar to scanning electron microscopy, a highly focused beam of ultrasound is transmitted and received by a single transducer. The generated signal is reflected by the concrete surface which is thoroughly scanned to obtain a point-by-point acoustical map. This method is especially suitable for identifying the location, orientation and width of cracks or pits in the surface of concrete.

Leaky Rayleigh Wave Technique

After the acoustic microscopy, this technique can be used to determine the depth of cracks or pits. Here an underwater source for Rayleigh waves and a separate underwater receiving transducer are used. When Rayleigh waves are directed towards the concrete surface, they radiate back into water unless they encounter a crack or a pit in the surface. The presence of cracks or pits in the concrete surface disturbs the Rayleigh waves and causes them to travel along the surface of the flaw, thus allowing the depth of the crack or the pit to be measured.

Rebar Location

Embedded metals in concrete can be located by a pachometer which is a metal detector that measures the magnetic field at the concrete surface. The pachometer can determine the depth and size of reinforcing bars, but it provides no information about corrosion of steel.

Rebar Corrosion

Dawson[6] has given an excellent review of electrochemical methods available to assess and monitor the corrosion of steel in concrete. ASTM Standard Method C 876 describes measurement of electrical potential using a reference electrode placed on the concrete surface, which is connected via a high-input impedance voltmeter to the reinforcement mat. Using $Cu/CuSO_4$ as a reference electrode, if the corrosion potential is greater than -0.20, the probability of corrosion is less than 5%; if the corrosion potential is less than -0.35, the probability of corrosion is more than 95%.

From extensive field tests in Denmark it was concluded that instead of a $Cu/CuSO_4$ reference electrode, the calomel electrode is more suitable for testing the corrosion of steel in concrete. Since the ASTM test for active corrosion from a $Cu/CuSO_4$ reference electrode is not considered reliable, it is suggested that the gradients of the isocorrosion potential contours (and where the contours branch together) provide a more reliable indication of a corroding or a corrosion-risk area.

According to Dawson, electrochemical noise measurement could be a valuable aid to the engineer since it provides quantification of the corrosion process and an indication of pitting corrosion. Using noise as amplitude, the spectral density plots were found to be useful for corrosion rate monitoring.

Linear polarization resistance measurement (LPRM) is the traditional DC technique for measuring corrosion rates in moist materials using the Stern–Geary equation. Clear[7] has described the field testing equipment, procedure, and experience with LPRM. A linear relationship is described mathematically for a region on the polarization curve in which slight changes in current applied to the corroding metal in an ionic solution cause corresponding changes in the potential of the metal. Simply stated, if a large current is required to change the potential to a given amount, the corrosion rate is high; whereas, if only a small current is required, the

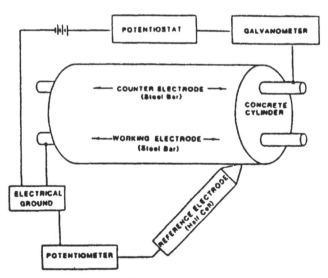

Fig. 8.3. Schematic of a linear polarization test set-up for a laboratory specimen (reproduced, with permission from Ref. 7).

Table 8.1. Guidelines for Data Interpretation.

I_{corr}	Corrosion damage
Less than 0·20 mA/ft²	No corrosion damage expected
0·20 to 1·0 mA/ft²	Corrosion damage in 10 to 15 years
1·0 to 10 mA/ft²	Corrosion damage in 2 to 10 years
More than 10 mA/ft²	Corrosion damage in 2 years or less

corrosion rate is low. A schematic of a LPRM test set-up by Clear[7] is shown in Fig. 8.3. The author recommends the guidelines given in Table 8.1 for data interpretation.

Core Tests

The above-described *in situ* nondestructive methods provide an inexpensive way to obtain a considerable amount of preliminary information on the condition of concrete. When these tests indicate internal deterioration, microcracking or zones of weakness in a concrete member, it is essential to perform direct tests on core specimens carefully obtained using a rotary diamond drill (ASTM C 42 Standard Method). A comparison between the core strengths and the original specified strength provides a reliable method of assessing the degree of deterioration in concrete.

Usually, petrographic analysis of concrete from core specimens is required for determining the probable causes of distress, such as lack of adequate air entrainment in concrete exposed to freezing and thawing cycles. Mineralogical analysis of the cement paste using optical microscopy or X-ray diffraction is essential to identify the chemical causes of deterioration, such as carbonation, sulfate attack, and alkali–aggregate reaction. In the case of corrosion of reinforcing steel, core specimens can be utilized to determine the distribution of chloride with respect to depth of cover from the concrete surface. The critical chloride level at the reinforcement surface to cause loss of passivity is assumed to be 0·4% chloride by weight of cement. Actually, it is the Cl^-/OH^- ratio rather than the chloride content which is now believed to be the critical factor in the corrosion of steel in concrete, and this underscores the importance of controlling the alkalinity of concrete for protecting steel from corrosion. Again, core samples are essential for identification of the causes of corrosion.

REPAIR MATERIALS

The selection of materials is more difficult for concrete repair than for new construction. The range of materials used is broader than cementitious materials because many polymeric compositions can be used effectively for concrete repair. Also, to insure durability of the repaired structure, the compatibility of various materials with each other, and with the original concrete, will have to be taken into consideration. Frequently, the speed at which the structure can be brought back into service is also an important consideration in the selection of materials and methods for repair.

Inorganic Cementitious Materials

Cementitious grouts, mortars, and concretes containing portland cement, high-alumina cement, sodium silicate cement, or calcined-gypsum cement are commonly used for repair of cracked and deteriorated concrete structures. They are readily available almost anywhere, relatively inexpensive, safe and easy to handle, and generally compatible with old concrete. Their main disadvantages are high drying shrinkage and cracking tendency, slow setting and strength development rates, and sometimes poor bonding properties.

Portland cement-based compositions are chemically similar to concrete and are, therefore, most compatible. When well bonded, their thermal compatibility to old concrete insures a stress-free interface which is essential to prevent delamination of the repair material. To obtain rapid setting of portland cement-based compositions, it is customary to use 2% to 4% calcium chloride by weight of cement when corrosion of steel is not a consideration (i.e. with unreinforced concrete structures). Sodium silicate, sodium aluminate, high alumina cement, and calcined gypsum can also be used as portland cement admixtures (10% to 15% by weight), when rapid setting and high early strengths are desired.

Mortars and concretes containing either high-alumina cement or calcined gypsum (plaster of Paris) can yield very high early strengths of the order of 60 to 70 MPa in one day; but high-alumina cement products undergo gradual strength loss especially under warm and humid conditions. Similarly, calcined gypsum products show considerable loss of strength on prolonged exposure to moisture. Cements containing up to 90% calcined gypsum are slightly expansive, which is a useful property for obtaining non-shrinking (crack-resistant) grouts and mortars for use in dry environments.

However, when resistance to moisture and cracking are simultaneously desired, it is not uncommon to recommend products containing Type K expansive cement. This cement is essentially a modified portand cement containing an anhydrous calcium sulfoaluminate ($4CaO \cdot 3Al_2O_3 \cdot SO_3$), which produces ettringite on hydration. The formation of ettringite results in expansion which, under restraining conditions, is manifested as prestress. The chemically induced prestress counteracts the tensile stresses caused by drying shrinkage, and thus reduces the chances of cracking.

Owing to their low permeability, high durability, high early and ultimate strength, and excellent bond strength, superplasticized portland cement mortar or concrete mixtures, with or without fly ash or silica fume, are becoming increasingly popular for repair and rehabilitation of concrete sea structures. They are especially suitable for application by shotcreting, and are described later. For underwater concrete repair, concrete mixtures containing anti-washout admixtures will also be described later.

Organic Materials

Organic materials such as surfactants, which are used in small dosages, are not discussed here. However, large amounts of polymeric resins have been used, either as a sole binder or in combination with portland cement, with considerable effect on setting, hardening, and permeability characteristics of the repair material. These materials, in spite of their high cost and toxicity,

offer distinct advantages in situations where the speed of repair
(e.g. a few minutes time of set as well as high compressive
strengths, of the order of 30–60 MPa, in 1 hour) is important, and
the areas to be repaired are rather small (e.g., filling of cracks or
construction of patches and overlays as thin as 5 mm).

Concretes containing polymers are classified into three categories. Polymer concrete (PC) is formed by *in situ* polymerization of
a mixture of a monomer and aggregate, with no other cementing
material present. The term, 'polymer impregnated concrete'
(PIC), is used when a hardened portland cement concrete is
impregnated with a monomer, which is subsequently polymerized
in situ. Latex modified concrete (LMC) consists of a conventional
portland cement concrete mixture, in which a part or all of the
mixing water has been replaced with a latex (polymer emulsion).
LMC hardens by air curing and does not develop the high
strengths and the high abrasion-resistance characteristics that are
typical of PC and PIC. However, LMC possesses excellent
bonding ability and impermeability and has been extensively used
in overlays for deteriorated bridge decks and industrial floors.
Application of PIC is limited to precast concrete products and
repair of concrete flatwork. Typical mechanical properties of PC,
LMC, and PIC are listed in Table 8.2. PC is most versatile for all
kinds of concrete repair work. Methyl methacrylate, polyester
styrene, urethane, and epoxies are the resinous materials usually
used for general repair and for filling of cracks, and will therefore
be discussed in more detail.

Methyl Methacrylates (MMA)
According to a report by Krause,[8] MMA monomer, when
suitably catalyzed for polymerization, provides 30 to 60 minutes
working time and then hardens rapidly to give 30 MPa compressive strength in 3 hours. At 24 hours, the typical compressive
and flexural strengths are 60 MPa and 15 MPa, respectively. The
use of a benzoyl peroxide catalyst and amine promoters is
common in PC formulations, and products with higher strengths
have been obtained by adding to the monomer system a silane
coupling agent which increases the interfacial bond between
polymer and aggregates.

Table 8.2. Typical Mechanical Properties of Concretes Containing Polymers (PSI).

	PC		LMC			PIC	
	Polyester	Polymerized MMA	Control		LMC containing styrene-butadiene air-cured	Control unimpregnated	MMA impregnated, thermal–catalytical polymerization
			Moist-cured	Air-cured			
polymer/aggregate ratio	1:10	1:15					
Compressive strength	18 000	20 000	5 800	4 500	4 800	5 300	18 000
Tensile strength	2 000	1 500	535	310	620	420	1 500
Flexural strength	5 000	3 000	1 070	610	1 430	740	2 300
Elastic modulus, $\times 10^6$	5	5·5	3·4	—	1·56	3·5	6·2

Source: PIC from ACI SP-40, LMC from ACI Committee 548, and PC from Dikeau, in Progress in Concrete Technology, ed. V. M. Malhotra, CANMET, 1980.

Before mixing with the monomer the aggregates must be completely dry, and the mixture bonds well to a clean and dry substrate. The major limitation of acrylic polymer systems is their inability to bond to wet or damp surfaces. The repair patch possesses excellent abrasion resistance and chemical resistance. In addition to cost, major disadvantages are flammability and toxicity (extended exposure may cause nausea and dizziness). According to Krause, a new high-molecular-weight MMA has been developed, which is less volatile and less flammable than the conventional MMA monomer.

Polyester–Styrene

Styrene monomer is used as a solvent to decrease the viscosity of unsaturated polyester monomer, and the solution is catalyzed to undergo copolymerization. Polyester resins are relatively less expensive: in 1985 the material cost for producing one cubic ft ($0.028\,m^3$) polyester–styrene PC was reported to be US $16, compared to $100 for MMA and epoxy resins. Commercial products are available with a wide variety of formulations, some capable of hardening to 100 MPa compressive strength at ordinary temperature within a few minutes of mixing the polymerizing catalyst. As with MMA concrete, it is reported that the use of a 0.5% to 1.0% silane coupling agent enhances the strength of a PC concrete containing siliceous aggregate and also increases the bond strength with old concrete substrates. Again, the monomers are volatile and flammable, and should be used with proper safety precautions.

Urethane

According to Krause,[8] urethane monomer possesses very low viscosity and is excellent for emergency repair work because it can be formulated to set within a few seconds. Aggregate is preplaced (usually, fines are omitted) in the cracked or spalled area and the urethane resin is flooded into the cavity using a two-component mixing gun (one component contains the hardener). Urethane patching materials can be used in cold or wet conditions and require minimum surface preparation.

Epoxy Resins

Epoxy resins are cyclic ethers, such as oxacyclopropanes, which normally harden by polymerization at room temperature. These are the most expensive and yet the most used among all the organic adhesives for concrete sealing, patching, and bonding. This is primarily because of their ability to adhere well to any type of surface, whether dry or wet. They develop high strength, show very low shrinkage and creep, and possess excellent resistance to abrasion, chloride penetration, and chemical attack.

Generally, epoxy resins are highly viscous (epoxy mortars may show putty-like consistency) and relatively slow-setting, and therefore continue to gain strength for weeks. Curing at very low or high temperatures requires special attention. With suitable chemical formulation, cure rates at room temperature can be varied from several minutes to several days.

Whereas epoxy coatings for surface protection and epoxy injection for bonding cracked concrete have generally enjoyed a good reputation, epoxy mortars and concrete patches have had mixed success in the past, mainly because of thermal incompatibility of the repair material with the substrate. This limitation has been largely overcome by the choice of appropriate aggregate type, and by careful manipulation of the aggregate grading and content. Thermal compatibility of a polymer concrete with the substrate can be improved by selecting an aggregate type that has a coefficient of thermal expansion similar to the aggregate used in the substrate. By using a well-graded aggregate, the epoxy content in the polymer concrete can be reduced to a minimum, which has the effect of improving the thermal compatibility further. Also, it should be noted that the greater the thickness of the repair patch, the greater will be the curing and thermal shrinkage stresses. The patch thickness and associated stresses should not exceed the strength of the polymer concrete.

The viscoelastic nature of epoxy polymers makes their mechanical behavior dependent on the glass transition temperature (T_g), defined as the temperature at which the polymer changes from a glassy (brittle) to a leathery (flexible) state. A value closely related to the T_g is the heat deflection temperature (HDT), defined as the temperature at which an epoxy beam deflects 0·25 mm (0·01 inch) under a constant stress of 1·82 MPa (ASTM D648). The HDT of an epoxy is important because all physical and mechanical properties including the coefficient of thermal expansion, bond strength, and elastic modulus undergo a drastic change at this temperature, which may be as low as 10°C for some epoxy formulations. The response of a viscoelastic material to loading is highly time-dependent and temperature-dependent. Therefore, to cope with creep in epoxies, it is important to use formulations with a high HDT and to distribute loads over large areas. Because there is a wide variety of commercially marketed formulations, a laboratory testing program on the rate of strength gain, bonding strength, heat deflection temperature, elastic modulus, thermal behavior, and moisture sensitivity should be undertaken before executing a major repair project involving epoxy resins.

CRITERIA FOR SELECTING REPAIR MATERIALS

The durability of a concrete repair depends both on the careful selection of the repair material and the method of repairing. A review of the important material properties that should be considered when selecting an appropriate repair material is given here. Experience shows that most repairs fail, not through lack of strength or lack of adhesion to old concrete, but through debonding of the repair material from stresses at the interface caused by drying shrinkage, differential thermal strain, and elastic mismatch. Accordingly, the following criteria for selecting repair materials are recommended:

Drying Shrinkage
Drying shrinkage cracking in cement concrete construction is controlled by providing reinforcing steel and construction joints. Most of the drying shrinkage is generally over within one year of construction; hence, old concrete is essentially free from drying shrinkage. Unless the cementitious material used for a repair patch is also shrinkage-free, the differential drying shrinkage strains will result in high stress at the interface between the old and the new

concretes, which is the major cause of cracking and debonding of repair patches.

Expansive additives, such as the ASTM Type K cement, which expands on hydration, provide the best means for combating the drying shrinkage cracking problem. Since unrestrained expansion results in loss of strength, such additives must be used under adequate restraining condition. Normal precautions to reduce the drying shrinkage should be taken, such as the use of a superplasticizer to reduce the water content of the repair mortar or grout. Also, there are construction methods that reduce the shrinkage potential. Examples of these methods are preplaced-aggregate concrete, dry-mix shotcrete, and dry-pack concrete.

Thermal Strain

All materials shrink on cooling and expand on heating, the thermal strain being equal to the coefficient of thermal expansion times the temperature change. Thermal shrinkage under restraint produces tensile stress which can cause cracking if the tensile strength of the material is low. Therefore, in environments subject to frequent thermal cycles, it is essential that the coefficient of thermal expansion of the repair patch or overlay is similar to the coefficient of thermal expansion of the concrete being replaced. Both with polymer concrete and portland cement concrete, one way to reduce the mismatch in thermal strains and elastic properties between the repair material and the substrate is to pay close attention to the aggregate in the repair mixture. The aggregate content of the repair mixture should be as high as possible, and the aggregate type should be the same or similar to the aggregate used in old concrete (i.e. has a similar elastic modulus and coefficient of thermal expansion).

Modulus of Elasticity

Materials with a low modulus of elasticity deform more when subjected to loading than materials with a high modulus of elasticity. When the external load is perpendicular to the bond line, the higher deformation of the lower-modulus material will transfer the load to the higher-modulus material, which may fail

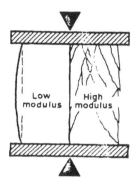

Fig. 8.4. Diagrammatic representation of failure of a repaired structure due to elastic modulus mismatch between the original material and the repair material (reproduced, with permission, from Ref. 9).

due to overloading. According to Warner,[9] this type of problem is likely to occur at the edge of a patch (Fig. 8.4), particularly when dynamic loading conditions are present (*impact or vibratory loads*), for instance in pedestals under large machinery or in structures exposed to wind or earthquakes. Repair mortars, concretes, and grouts of significantly higher elastic modulus than the substrate are undesirable not only from potential failure under externally applied loads, but also from internally induced stresses caused by drying shrinkage and thermal shrinkage strains, as discussed earlier.

Permeability
Good quality concrete is impermeable to liquids but is able to 'breathe' (i.e. it is permeable to water vapor). If a material used for patching, overlays, or coating of a concrete slab on grade (or any structure which is exposed to water on one side) is so impermeable that it is unable to 'breathe', the entrapped water vapor can cause high pressures and consequently a failure of the repair material either at the bond line or within the material. Solvent-based asphalt, methyl methacrylate, and epoxy materials are 'non-breathers' and frequently fail by debonding when used on moist concrete slabs or walls. 'Breather' materials include portland cement products, latex emulsions, and water-based epoxies.

According to Warner,[9] impermeable materials should also be avoided in patching concrete that has been damaged owing to corrosion of the reinforcing steel. Electric conductivity of concrete is one of the factors affecting corrosion, and all concrete, unless it is completely dry, is conductive. Should a portion of the concrete be replaced with a nonconductive repair patch, the current will be concentrated in a smaller area and the rate of corrosion in this area will be accelerated. Warner cites field experience with a reinforced concrete beam which was repaired after damage by corrosion cracking and spalling, using a nonconductive mortar. Within a year severe corrosion of the rebar was evident, even though the steel had been cleared of all corrosion product during the repair. The repair work was of good quality in every respect except for the use of a nonconductive material. Because of this negligence in the material selection, the entire structure had to be demolished and replaced only a year after extensive repairs were made.

EXECUTION OF REPAIR

The process of executing the repair work begins with the selection of appropriate methods of removing the deteriorated concrete and preparing the surface for good adhesion, and ends with placement of the repair mortar or concrete. Based on experience gained on a number of rehabilitation projects, Saucier[10] concluded that concrete work on repair projects requires much greater attention to good practice than may be necessary for new construction. Procedures not ordinarily considered in new construction, such as shotcreting, trowelling, and application by hand have to be considered for repair work. In this section, an overview of the common practices used in various steps of the repair process is given. Specially reviewed are shotcreting, epoxy injection, protective coatings, underwater repair, and repairs of reinforced concrete damaged by chloride corrosion.

Methods of Removing the Deteriorated Concrete

Concrete that is cracked but otherwise strong and sound can be repaired by epoxy injection, usually without the need for demolition of the cracked member. Any deteriorated, unsound, and loose concrete must be removed before applying the repair patch. Selected concrete removal techniques (note that the same removal technique may not be suitable for all parts of the structure) should be safe, economical, and time-saving. Saucier[10] has classified as follows the concrete removal methods by the manner in which the removal process acts on concrete.

Blasting

Blasting methods generally employ rapidly expanding gas confined within a series of boreholes to produce controlled fracture in concrete. Explosive blasting is the traditional technique which involves placing an explosive in each core hole and detonating the explosive. The high-pressure carbon dioxide blaster employs the carbon dioxide gas to break down the material. Wet and dry sand-blasting are also commonly used for removal of small amounts of concrete.

Cutting

Cutting methods generally depend on mechanical sawing, intense heat application, or very high-pressure water jets, 10–35 ksi (approx. 70–200 MPa), to cut around the perimeter of concrete sections in order to allow their removal. The equipment used for *water-jet cutting* is expensive and may be uneconomical unless large volumes of concrete are to be removed. According to Cambell,[11] the technique is attractive for concrete deck rehabilitation because reinforcing steel is not damaged and new microcracks are not produced in concrete. It is especially suitable for repair of navigation locks because dewatering of the lock is not essential, which reduces the overall rehabilitation time and cost. In one case the underwater concrete removal rate was found to be less than the above-water removal rate, probably because the underwater concrete is generally stronger.

The basic principles and equipment associated with water-jet cutting are described by Guha.[12] According to the author, when water is compressed to an ultra-high pressure level and released through a small opening, the expanding water can attain an estimated speed up to three times the velocity of sound. The focused jet of water traveling, for instance at 870 m/s, releases sufficient kinetic energy to cut through most hard materials. For many commercial environments a significant performance criterion is the cutting speed. At a given kinetic energy level a water jet will eventually cut through a material of a certain thickness and hardness. However, to improve the cutting speed to a commercially viable point, the mass of the water may be increased by adding fine particles of an abrasive mineral, such as silica or alumina. With regard to equipment, the important components are an intensifier (oil under hydraulic pressure is used here for the conversion of line-pressure water into ultra-high pressure water), an accumulator, the cutting nozzle, and water treatment and filtration units.

Impacting

Impacting methods, such as jackhammering, involve repeated striking of a concrete surface with a tool to fracture and break the concrete.

Presplitting

Blasting and impacting methods may be considered destructive for repair of complex concrete structures. The deteriorated portion can be separated from the structure by less destructive presplitting methods, such as mechanical wedging, water jetting, or the use of an expansive cement in holes drilled along the desired crack plane.

Surface Preparation

Methods of removing the deteriorated concrete may not leave a surface clean enough for good adhesion by the repair material. Sometimes concrete removal is not needed; however, the surface must be made free from laitance, oil, grease, or old paint. Since good bonding between the old and the new concrete is essential for durability of the repaired structure, surface preparation is as necessary as the thorough removal of bad or unsound concrete.

Surfaces damaged by erosion/abrasion are generally smooth and should therefore be roughened to enhance bonding to new concrete. High-pressure water jetting or light impacting tools can be used for this purpose. With properly prepared surfaces, usually no bonding agents are needed. In areas of high sedimentation or active marine growth for underwater repair, the placement of new concrete should be carried out promptly after the surface preparation.

Murray[13] has given a brief review of surface preparation methods. The mechanical methods include dry or wet sand-blasting, shot-blasting, bush-hammering, surface-scraping with a carbide-tipped wheel, and grinding. The chemical cleaning methods involve removal of contaminants by application of detergents, TSP (trisodium phosphate), or other proprietary mixtures, followed by surface etching with dilute hydrochloric acid. The hydrochloric acid, commonly referred to as muriatic acid, dissolves the cement paste and exposes the aggregate for bonding with the repair material.

An excellent bond of fresh mortar or concrete to old concrete can be achieved with proper surface preparation and without the use of any bonding agents. However, with thin patches (less than 50 mm thick), a bonding coat is generally recommended before placement of the repair material.[10] Mixtures of portland cement and fine sand with a latex emulsion make excellent bonding agents.

Some sort of formwork is usually needed to mold the repair material into a specific shape and size. The formwork may be removed after the repair material has hardened, or it may be left in place, such as steel forms designed to protect concrete from abrasion in the splashing zone. Rigid forms made of timber, plywood, or fiber-glass are used primarily to repair flat surfaces such as decks, foundations, and sea walls. A jacket is usually a free-standing, thin-walled, cylindrical element used primarily for the repair of pilings and caissons, and may be made from a variety of materials including nylon, cardboard, plywood, and fiber-glass. The jackets are generally fabricated as two half-cylinders which are centered around the pile with the help of spacer blocks and are bolted together. After the repair material has hardened, the form is usually removed.

Shotcreting

Where thin repair sections (e.g., less than 150 mm thickness) having large surface areas with irregular contours are needed, installation of the repair material by shotcreting may be more economical than conventional concreting because of savings in the forming cost. With the advent of high-strength and cohesive concrete mixtures, such as those containing superplasticizers and condensed silica fume or fly ash, the wet-mix shotcreting process is increasingly being used to repair deteriorated concrete seawalls, piles, and other hydraulic structures.

According to Heneghan,[14] latex-modified shotcrete was used to repair the heavily spalled beams of the Roosevelt Boulevard Bridge over the Intercoastal Waterway in Ocean City, New Jersey, and the pile caps of the Sunshine Skyway Bridge in St Petersburg, Florida, which had been damaged by salt water corrosion. The author also described the materials and procedure used for the shotcrete repairs to a portion of a pier in Portland, Maine, in 1977.

A major wet-mix shotcreting repair project, carried out in 1986 in St John, New Brunswick, is described by Gilbride et al.[15] The wharves in St John are subject to an unusually high tidal wave (8·5 m) and over 200 freezing and thawing cycles a year. The deteriorated and spalled concrete were repaired using steel-fiber-reinforced, silica-fume shotcrete anchored with grouted dowels.

Morgan[16] describes his satisfactory experience with over fifty different structures that have been recently repaired using wet-mix shotcrete systems containing superplasticized concrete with silica fume. According to the author, compared to the conventional dry-mix method, the wet-mix method is a significant improvement for *repair of vertical faces*. In dry-mix shotcreting the single-pass thickness of an overhead spray is limited to 2–3 in. (50–75 mm), and the use of set accelerators is generally detrimental from the standpoint of durability. For overhead applications, using wet-mix silica fume concrete, up to 18 in (450 mm) thick material can be applied by shotcreting without sloughing or tearing. Thus, it is claimed that sea walls up to several feet thick on vertical faces can be rebuilt without any problems.

The shotcreting technology has also emerged as a solution to overcome the difficulty of *repairing structures in the tidal zone.* Morgan[16] described shotcreting the face of a bulkhead wharf structure from a barge which rose and fell with the 8 m high tide (the barge dropped at the rate of 1·2 m an hour). Shotcrete, up to 9 in (225 mm) thick, was applied on the falling tide. No accelerators were used and the concrete did not set for 6 to 8 hours in cold temperatures, yet no shotcrete was lost by the tidal action.

Frost-resistant shotcrete can be obtained by using an air-entrained mixture and a funnel gun in the wet-mix process. According to Saucier,[17] the funnel gun is a pneumatic apparatus for placement of relatively small volumes of a stiff mortar, and therefore is suitable for repair work.

Shotcreting by the wet-mix process is also useful when fiber-reinforced concrete mixtures have to be applied. Fiber-reinforcement of concrete makes the material highly resistant to repeated impact and flexural fatigue, the properties which are of concern in the splashing zone of marine structures. Ramakrishnan et al.[18] evaluated the comparative fatigue resistance for concrete mixtures containing 0·5% and 1·05% by volume using four types of fibers, namely hooked-end steel fiber, straight steel fiber, corrugated steel fiber, and polypropylene fiber. The addition of the four types of fibers caused a considerable increase in the flexural fatigue strength and the endurance limit for 4 million cycles, with the hooked-end steel fibers providing the highest improvement (143%) and the straight steel and polypropylene fibers providing the least. The impact strength was increased substantially by the addition of all four types of steel fibers.

Morgan et al.[19] also presented the results of a comparative evaluation of toughness index and residual load-carrying capacity after first crack for wet-mix shotcretes containing steel fibers and plain or collated fibrillated polypropylene (CFP) fibers. It was concluded that the incorporation of relatively higher volume concentrations than normal (e.g., $6 kg/m^3$) of 38-mm CFP fiber in wet-mix shotcrete presents opportunities for a wide range of applications where a tough, ductile, corrosion-resistant material is required. Thus CFP fiber is considered to be a viable alternative to traditional mesh or steel fiber reinforcement of wet-mix shotcretes for many applications.

Epoxy Injection

Epoxy injection has been used successfully in the repair of cracks in caissons, piles, decks, sea walls, and foundations. Cracks as narrow as 0·002 in (0·05 mm) can be bonded by the use of the epoxy injection technique. Saucier[10] warns that unless the crack is stable or dormant (or the cause of cracking has been removed, thereby making the crack dormant), it will probably recur, possibly somewhere else in the structure. The general steps in the epoxy injection repair process are as follows:

(1) *Cleaning the cracks and surface sealing*: Before injection, surface cracks should be sealed to keep the epoxy from leaking out before it has hardened.

(2) *Installing the entry ports*: At close intervals along the cracks, holes have to be drilled and usually entry ports for epoxy injection have to be installed.

(3) *Injecting the epoxy*: Hydraulic pumps, pressure pots, or air-actuated caulking guns can be used for injecting the epoxy mixture under pressure. The use of excessive pressure can propagate the existing cracks and is, therefore, to be avoided. Also, care must be taken to prepare the amount of adhesive mixture that can be used safely before the commencement of hardening during the injection process.

Snow[20] has illustrated the various phases of the pile encapsulation process, using a fiber-reinforced plastic jacket and epoxy grout injection (see Fig. 8.5). His paper contains an outline of factors that influence the long-term durability of polymer-encapsulation systems. The most prevalent problems causing the lack of durability include discontinuity of polymer grout, lack of bond between the polymer grout and the plastic jacket, lack of bond between polymer grout and substrate, improper mixing or curing of grout, thermal incompatibility, and ultraviolet deterioration of the jacket material. The author has also described methods of *in situ* testing and job-site monitoring.

Protective Coatings and Linings

Repaired concrete and even new concrete members are frequently provided with additional protection from the physical–chemical

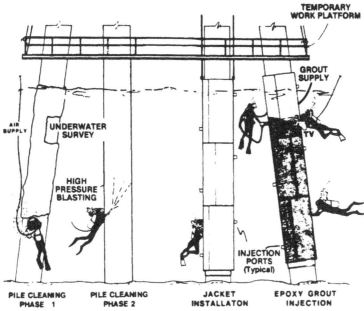

Fig. 8.5. Various phases of a successful polymer encapsulation process are shown from left to right (reproduced, with permission, from Ref. 20).

processes of deterioration by the use of surface coatings and linings. The approach underlying the concrete surface protection strategies is either to reduce the absorption of water (i.e. to keep the concrete dry because without water no degradation can occur) or to prevent contact between the concrete surface and the aggressive environment (such as seawater). The first approach involves the use of either hydrophobic agents which do not react with the constituents of concrete or certain reactive agents which may fill up the pores with reaction products. The second approach depends on the application of surface linings, such as prefabricated polymer membranes, rubber linings, sheets of plastic or stainless steel, and ceramic tiles or granite panels.

Based on an excellent survey of the subject by Bijen,[21] various types of surface protections are illustrated in Fig. 8.6 and summarized in Table 8.3. According to the author, the hydrophobic agents capable of imparting 10 years or more of water-repellence are generally organosilicon compounds of the following

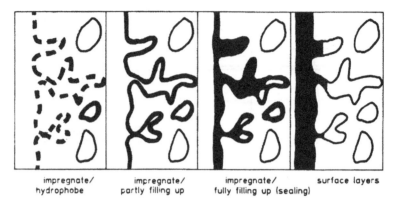

impregnate/   impregnate/   impregnate/   surface layers
hydrophobe   partly filling up   fully filling up (sealing)

Fig. 8.6. Types of surface protection (reproduced, with permission, from Ref. 21).

type:

- Siliconates with a molecular weight, $M = 100–200$.
- Silicon resins, $M > 2000$.
- Alkylalkoxy silanes, $M = 100–200$.
- Oligomeric alkylalkoxy siloxanes, $M = 400–600$.
- Polymeric alkylalkoxy siloxanes, $M > 1000$.

These agents can form chemical bonds with the constituents of concrete, with the nonpolar group providing the water-repellent properties. Although a large number of protective materials seem to meet the Kunzel criteria (see Fig. 8.7), Bijen[21] suggests that hydrophobic rendering of concrete is a better protective measure than the use of sealants or impermeable coatings. Problems with sealants and impermeable coatings can be avoided if the concrete is dried before it is protected, and if it does not get resaturated by capillary action. A critical problem with linings is the possibility of debonding, which occurs frequently when the substrate has not been well prepared.

In conclusion, unless one is thoroughly familiar with the application, properties, and limitations of coatings and linings, it is preferable not to depend on the protection provided by their use. With good-quality concrete, such protective devices are not necessary; with poor-quality concrete, the protection is at best temporary because it will depend on the durability of the coating or lining, which may not be very high.

Table 8.3. Survey of Various Methods of Surface Protection.

Treatment	Effect	Substrate conditions
Impregnate/ hydrophobe	Water-repelling/water vapor permeable; not resistant to chemical loading	Fine pores
Impregnate/ fill up pores	Decrease water absorption and increase water and water vapor resistance; not resistant to chemical loading	Fine pores
Thin coatings	Water and water vapor tight; sensitive to mechanical loading; restricted chemical and thermal resistance	Smooth surface, free from large pores and cracks[a]
Thick coatings	Water and water vapor tight; more resistant than thin coatings	Smooth surface, fine cracks allowed
Inorganic plasters	Fairly watertight, water vapor permeable; non-chemical resistant (excluding special types)	Free of large macropores (air bubbles, honeycombs) non-moving fine cracks allowed[a]
Organic plasters/ high built coatings	Water and water vapor tight; resistant to chemical loading, less resistant to mechanical loading	Free of macropores, non-moving fine cracks allowed[a]
Film membranes	Water and water vapor tight; resistant to chemical loading; less resistant to mechanical loading	Smooth surface, cracks to 3 mm width allowed
Rubber lining	Water and water vapor tight, resistant to chemical, temperature and mechanical loading	Smooth
Lining with thermoplasts, sheeting or pipes	Water and water vapor tight, resistant to chemical, temperature and mechanical loading	Smooth
Tiling	Resistance, depending on kind of tiles, adhesive and joint filler	Smooth

[a] *Crack bridging ability can be increased by fiber reinforcement.*
Reproduced, with permission, from Ref. 21.

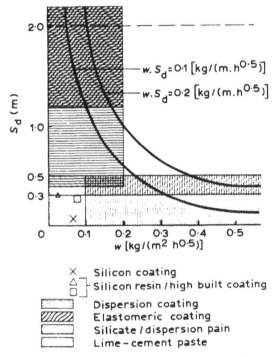

Fig. 8.7. Kunzel relation between water absorption coefficient (*W*) and water vapor diffusion resistance (*S*$_d$) and positions of various protective agents (reproduced, with permission, from Ref. 21).

Underwater Concreting

A number of techniques are available for underwater concrete placement. Although tremie is the most commonly used method with new construction, which usually requires relatively large quantities of concrete, it is not the preferred method for repair. To prevent dilution of the cement paste by water, the lower end of the tremie pipe must always remain embedded within the mass of fresh concrete. It is difficult to maintain this seal with the small amounts of concrete normally needed in repair. Therefore, for relatively shallow placements, either pumping or freely falling concrete (by gravity), are the two commonly used methods for underwater concrete repair.

Underwater concrete placement for repair by pumping does not differ significantly from placements for new construction, and compared to tremie the equipment for pumping is less elaborate.

The use of a pump with a boom makes it easy to place concrete at various points within a certain area. The concrete mixture should be cohesive, and to prevent dilution with water it is placed as close to the final resting site as possible.

A promising development is the use of *anti-washout admixtures* which permit the free dropping of concrete through water without the need for confinement, such as the tremie pipe or pump line. The addition of anti-washout admixtures makes the concrete mixture so cohesive that a free fall through water does not result in any significant loss of fines or in an increase in water/cement ratio. The admixtures consist of water-soluble natural gums or synthetic polymers, which increase the viscosity of the mixing water. Typical examples are cellulose ethers, pregelatinized starches, polyethylene oxides, alignates, and polyvinyl alcohol. The dosage, depending on the amount of filler materials in the product, ranges from 0·1% to 2·5% solid by weight of cement. Dosages above the optimum levels may cause excessive retardation and air entrapment, which will have an adverse effect on strength. Commercial products sometimes contain short fibers, 200–400 μm long, in addition to fillers, such as pulverized limestone or silica.

Hester *et al.*[22] described the results of a comprehensive laboratory investigation on selecting materials, mix proportions, and procedures for underwater concreting. Several types of water-reducing admixtures (normal and high range), mineral admixtures (fly ash, slag, and silica fume), and anti-washout admixtures were tested. The dosage of anti-washout admixture was adjusted to obtain a washout mass loss of less than 5%, after three drops through a standard column of water. Although both silica fume and anti-washout admixtures reduce the loss of fines by increasing the viscosity of water in the concrete mixture, their dosage has to be limited because of the excessive water requirement, which is harmful to the strength and abrasion resistance of concrete. For flowing concrete mixtures with optimum rheological and mechanical properties, the authors presented the following conclusions and recommendations:

> The cement content should be limited to approximately 350 kg/m³, with 5% to 12% fly ash addition for better workability, and about 7%

silica fume for strength enhancement. Using a superplasticizing admixture, flowing concrete can be obtained with approximately 0·4 water/cementitious ratio. An underwater placement test showed that the use of an anti-washout admixture, 0·35% by weight of cement for a high-purity product (and 1·25–2·5% for products containing fillers) limited the washout mass loss to a maximum of 3%. The use of a de-airing agent is recommended to keep the volume of entrapped air below 3%. For obtaining high resistance to abrasion, a hard pea-gravel, with a minimum coarse aggregate content of 40% by weight is recommended. The maximum fine aggregate (natural silica sand) content should be approximately 46% by weight of the total ag-gregate. In a 72-hour standard abrasion/erosion test the recommended concrete mixtures gave approximately 3–5% weight loss for above-water placement and 4–5% loss for underwater placement. The 7-day and 28-day compressive strengths were 35–40 MPa and 45–60 MPa, respectively.

To patch shallow (for instance, less than $\frac{1}{3}$ m) unreinforced scour holes in relatively fast-moving water or for placement of thin protective overlays on sloping or flat surfaces, Khayat[23] recom-mends rich concrete mixtures with 450 and 70 kg/m³ cement and silica fume contents respectively, and water/cementitious ratios not exceeding 0·35. For underwater repair of large scour holes ($\frac{1}{3}$ to 1 m deep), the cementitious materials in concrete should consist of 360 kg/m³ cement, 24 kg/m³ silica fume, and 18 kg/m³ fly ash, with a 0·4 water/cementitious ratio. For larger and deep scour holes with steel reinforcement, a somewhat higher water/ cementitious ratio (0·45) may be necessary to insure high fluidity. All concrete mixtures should contain superplasticizing as well as anti-washout admixtures because they should not only be self-compacting and self-leveling but also must develop high strength and high erosion resistance. Khayat was able to obtain repair surfaces with *in situ* compressive strength in excess of 55 MPa (8000 psi) and approximately 2 MPa bond strength between the underwater-cast concrete and the underlying concrete surfaces. According to the author, a small volume of steel fibers may be used to improve the toughness of the repair concrete without adversely affecting the abrasion resistance.

The Concrete Society's (United Kingdom) recommendations on underwater concreting are currently under review. A preview of these recommendations is presented by McLeish.[24] According to his report, the recommendations have taken into consideration the recent developments in materials and placement methods which have enhanced the quality of underwater-placed concrete and have enabled the placement of thinner and more complex sections that are particularly suitable for repair work.

REPAIR OF REINFORCED CONCRETE DAMAGED BY CHLORIDE CORROSION

The original cause of deterioration of a given reinforced concrete structure exposed to a marine environment may lie in microcracking due to one or more of the numerous physical–chemical factors, such as thermal shock cycles of freezing and thawing or heating and cooling, and alkali–silica reaction. However, as discussed before, in the advanced stages of deterioration when repair is undertaken, the chloride-initiated corrosion from seawater is almost invariably present. It is, therefore, useful to review the various steps that are involved in repairing structures damaged by chloride-initiated corrosion.

Generally, when chloride ions have to penetrate into concrete from outside it is found that the corrosion of the reinforcement is related to the chloride concentrations; in other words, most severe corrosion of the reinforcement is usually found where the chloride ion concentration is the highest. An exception to this is the concrete in the fully submerged zone, where lack of oxygen will limit the corrosion. Probably for the same reason, the rate of chloride-initiated corrosion is also found to be dependent on the quality of the concrete cover. According to Bijen,[21] in the Federal Republic of Germany preliminary regulations on the repair of concrete list limiting values for the chloride concentration. For a good-quality concrete a limiting value is stated to be 0·6% to 0·8% by mass of cement; for a low-quality concrete, it is 0·3% to 0·4%. This means that in every case it may not be essential to remove concrete with more than 0·3–0·4% chloride content unless evidence of steel corrosion is present.

Browne *et al.*[25] have proposed a comprehensive approach for diagnosis and repair of reinforced concrete marine structures contaminated with chloride. A range of *in situ* tests is used to evaluate the extent of deterioration in damaged areas and the risk of future deterioration in hitherto undamaged areas. The service life of individual elements of a structure can be predicted from empirical relationships involving cover thickness, concrete quality, and the chloride level in concrete. According to the authors, from a combination of nondestructive tests, *in situ* monitoring of the structure, and core tests, a specification for repair can be written which allows for variations in the severity of chloride corrosion attack.

Two methods are available for repairing the structures affected by chloride corrosion. The traditional method involves removal of chloride-containing concrete, insulating the reinforcement, and installation of repair mortar or concrete. The second method, namely the cathodic protection of steel in concrete, is being increasingly used especially when complete removal of chloride is

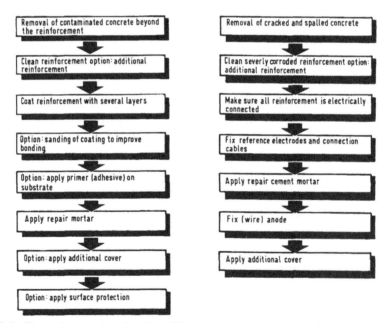

Fig. 8.8. Flow sheets showing 'traditional' repair procedure (left) and repair procedure with CP (right) for chloride-initiated corrosion (reproduced, with permission, from Ref. 21).

not possible for structural reasons. Various steps in both the methods are illustrated by Fig. 8.8, and will be discussed next.

The Traditional Repair Method

The first step in the repair process is the removal of chloride-contaminated concrete to at least 10 mm behind the reinforcement. (The distance should be increased to 15 or 20 mm for large diameter bars.) Bijen[21] recommends that the concrete must be removed to the point that adjacent to the corroded areas at least 100 mm of non-corroding reinforcement is revealed. When contaminated concrete is being removed, the structural safety of the structure during and immediately after repair must be safeguarded.

After removal of the concrete, the reinforcement has to be cleaned to remove corrosion products and corrosion pits. This is generally done by wire brushing or sand-blasting, which is not always effective in removal of deep pits. As stated before, cleaning by high-pressure water jetting appears to give better results. New or additional reinforcement may be installed for structural reasons.

The cleaned steel surface is next coated with an insulating coating to ensure that any chlorides remaining in the adjacent or old concrete would not renew the damage. To cover the defects, coatings are applied in two or more layers; the last coating may have to be sanded to provide better adhesion to the repair mortar or concrete. Epoxy coating products are most commonly used in field practice (e.g., repair of the spandrel beams of the San Mateo–Hayward Bridge), although primarily due to the lower cost styrene–butadiene latexes are also becoming popular, as will be discussed below. Compared to epoxy coatings, latex-modified cement coatings have the additional advantage of providing an alkaline environment at the reinforcement.

Before applying the repair mortars or concrete, an insulating adhesive as a primer is often used to improve the adhesion of the repair material to the substrate and to insulate the reinforcement from the repair material electrically. Insulating the reinforcement by drying the old concrete before installation of repair mortar may also be considered, but complete drying of chloride-contaminated concrete, and keeping it dry during service, is difficult (owing to the hygroscopic character of salts of chloride).

Finally, a carefully selected repair mortar or concrete is installed to a predetermined cover thickness. The properties, advantages, and disadvantages of various repair materials have been discussed earlier. When the damage involves internal chloride contamination from the use of unwashed aggregate in concrete or chloride-containing admixtures and there are no reasons to expect chloride penetration from outside, the repair is regarded as complete. In the case of marine structures, the danger of external chloride penetration can never be completely ruled out. Therefore, after the repair it is desirable to consider the option of a chloride-resisting surface protection.

With regard to cover thickness, according to Bijen[21] the advantage of extra cover thickness is that not only will it provide additional protection to concrete against further chloride penetration from the exterior, but also it will help in levelling out the existing chloride concentration gradients (Fig. 8.9), which play an important part in setting up the corrosion cells.

The AR (Asano Refresh) process,[26] a repair method developed by a Japanese cement company for salt-damaged reinforced concrete structures, utilizes a corrosion-inhibitor coating on the reinforcement, a latex mortar for repair, and an organic paint to protect the surface of the repair mortar. The outline of the process

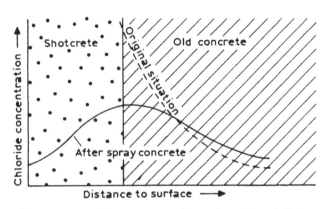

Fig. 8.9. Chloride concentration expressed as a function of distance from the surface before and some time after applying sprayed concrete (reproduced, with permission, from Ref. 21).

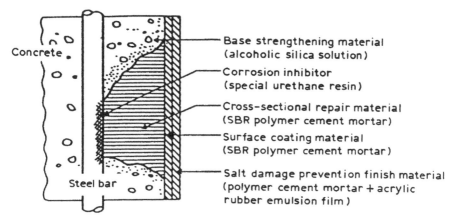

Fig. 8.10. Outline of the Asano Refresh process (reproduced, with permission, from Ref. 26).

is shown in Fig. 8.10. Prepacked aggregate grouted with styrene–butadiene latex mortar is used as the principal repair material. Proper formwork has to be provided to support the prepacked aggregate. After removal of the formwork, a surface coat consisting of styrene–butadiene and acrylic rubber latexes and incorporating a layer of alkali-resistant glass-fiber mesh helps to protect the repaired structural element by preventing the ingress of water, chloride, and oxygen. An experimental study[27] showed that a thickness of 3 mm or less of the latex concrete did not have much effect on the water transmission rate, but a thickness of 10 mm gave a water transmission rate of less than 1 mg/cm² day. Using a 12-mm thickness of styrene–butadiene latex concrete, the chloride penetration was completely inhibited. A combination of polymer concrete and acrylic rubber emulsion surface coating provided an excellent salt-damage prevention system, but neither the polymer concrete nor the paint filler alone were as effective.

Repair Methods Involving Cathodic Protection of Steel

For a variety of reasons, cathodic protection of steel in concrete is being increasingly used for repair of chloride-damaged concrete structures. First, for structural reasons or owing to congestion it may not be feasible to remove the chloride-contaminated concrete.

However, for satisfactory repair, it is required that the repair mortar must fully envelop this reinforcement. Second, sometimes it is difficult to get rid of all the corrosion pits by wire brushing or sand-blasting. Third, the repaired areas start acting as cathodes, and the unrepaired areas become anodes and start corroding if sufficient oxygen and electrolytic conductance are present.

Among others, Clear[7] has described in detail the cathodic protection techniques. The underlying principle involves impressed-potential from an outside source to artificially lower the electrical potential of the corroding reinforcement. This diminishes the current density and thus the rate of the corrosive anodic reaction. The primary components of the cathodic protection system are a source of DC current, an anode system, a conductive concrete (moist concrete is conductive), a cathodic system (the steel reinforcement), electrical connecting cables, and control devices (such as a reference electrode and data acquisition system). A conductive overlay usually forms the anode.

Fig. 8.8 illustrates the similarities and differences between the traditional repair method and the method for repair based on cathodic protection. Briefly, the installation of the cathodic protection involves the following steps: removal of cracked concrete and filling up of cracked or spalled areas with ordinary cement mortar (or concrete) which is electrically conductive. Severely corroded reinforcement should either be cleaned or replaced before filling up with repair mortar. The repair mortar surface is sand-blasted to provide a good substrate for a conductive overlay (such as asphalt with coke breeze as a filler). Subsequently, the conductive overlay (anode) is installed. After completion the cathodic protection system is checked periodically, and adjustments, if needed, are made in the impressed voltage and current.

EVALUATION OF REPAIR SYSTEMS

The science and engineering of concrete repair are progressing rapidly; however, there is not much information in the published literature on evaluation of different repair systems. Coote *et al.*[27] describe an investigation involving small reinforced concrete specimens that were repaired either with systems containing ordinary portland cement or systems containing epoxy resin. The specimens were exposed to the tidal zone in a marine environment over periods up to $6\frac{1}{2}$ years. So far the results have indicated that portland cement-based repair formulations are better than epoxy-based formulations in providing corrosion protection to reinforcing steel and preventing general deterioration of concrete.

REFERENCES

1. Khanna, J., Seabrook, P., Gerwick, B. C. & Bickley, J., Investigation of distress in precast concrete piles at Rodney Terminal, *Performance of Concrete in Marine Environment,* ed. V. M. Malhotra, ACI SP-109, 1988, pp. 277–320.
2. Khanna, J., Gilbride, P. & Whitecomb, R., Steel fiber reinforced concrete jackets for repairing concrete piles, *Performance of Concrete in Marine Environment,* ed. V. M. Malhotra, ACI SP-109, 1988, pp. 227–52.
3. Mehta, P. K., Durability of concrete exposed to marine environment—a fresh look, *Performance of Concrete in Marine Environment,* ed. V. M. Malhotra, ACI SP-109, 1988, pp. 1–23.
4. Leendertse, W. & Oud, H. J. C., The Dutch experience with construction and repair of marine structures, *Proceedings, Gerwick Symposium on Durability of Concrete in Marine Environment,* ed. P. K. Mehta, Dept. of Civil Engineering, University of California at Berkeley, 1989, pp. 148–56.
5. Gerwick, B. C., Review of the state of the art for underwater repair of abrasion-resistance concrete, Report to the US Army Corps of Engineers, *REMR CS-19,* 1988.
6. Dawson, J. L., Corrosion monitoring of steel in concrete, *Corrosion of Reinforcement in Concrete Construction,* ed. A. P. Crane, Ellis Horwood, Chichester, 1983, pp. 175–92.
7. Clear, K. C., Cathodic protection of reinforced concrete members, *Proceedings, Gerwick Symposium on Durability of Concrete in Marine Environment,* ed. P. K. Mehta, Dept. of Civil Engineering, University of California at Berkeley, 1989, pp. 100–23.

8. Krause, P. D., New materials and techniques for rehabilitation of portland cement concrete, State of California, Dept. of Transportation, Report No. FHWA/CA/TL 85-16, 1985.

9. Warner, J., Selecting repair materials, *Concrete Construction*, October (1984), 865–871.

10. Saucier, K. L., Repair and rehabilitation of concrete structures, *Proceedings, Gerwick Symposium on Durability of Concrete in Marine Environment*, ed. P. K. Mehta, Dept. of Civil Engineering, University of California at Berkeley, 1989, pp. 173–80.

11. Cambell, R. L., Evaluation of water jet blasting for removal of concrete from lock chamber faces, *The REMR Bulletin* (US Army Corps Engineers), **6**(5) (1989), 4–7.

12. Guha, J. K., High-pressure waterjet cutting—an introduction, *Ceramic Bull.*, **69**(6) (1990), 1027–9.

13. Murray, M. A., Surface preparation for adhesives, *Concrete International*, **11**(9) (1989), 69–71.

14. Heneghan, J. I., Shotcrete repair of concrete structures in marine environment, *Performance of Concrete in Marine Environment*, ed. V. M. Malhotra. ACI SP-65, 1980, pp. 509–26.

15. Gilbride, P., Morgan, D. R. & Bremner, T. W., Deterioration and rehabilitation of berth faces in tidal zones at the Port of Saint John, *Performance of Concrete in Marine Environment*, ed V. M. Malhotra, ACI SP-109, 1988, pp. 199–225.

16. Morgan, D. R., Discussion of papers on concrete construction and repair, *Proceedings, Gerwick Symposium on Durability of Concrete in Marine Environment*, ed. P. K. Mehta, Dept. of Civil Engineering, University of California at Berkeley, 1989, pp. 185–6.

17. Saucier, K. L., Placing specialized concretes, *Concrete International*, **12**(6) (1990), 46–50.

18. Ramakrishnan, V., Wu, George, Y. & Hosalli, G., Flexural fatigue strength, endurance limit, and impact strength of fiber-reinforced concrete, *Transportation Research Record*, **1226** (1989), 17–24.

19. Morgan, D. R., McAskill, N., Richardson, B. W. & Zellers, R. C., A comparative evaluation of plain, polypropylene fiber, steel fiber, and wire mesh reinforced shotcretes, *Transportation Research Record*, **1226** (1989), 78–87.

20. Snow, R. K., Polymer pile encapsulation: factors influencing performance, *Concrete International*, **12**(5) (1990) 34–40.

21. Bijen, J. M., Maintenance and repair of concrete structures, *HERON* (Delft University of Technology, The Netherlands), **34**(2) (1989), 1–82.

22. Hester, W. T., Khayat, K. H. & Gerwick, B. C., Properties of concrete for thin underwater placements and repairs, *Fly Ash, Silica Fume, Slag, and Natural Pozzolans in Concrete,* ed. V. M. Malhotra, ACI SP-114, 1989, pp. 713–31.

23. Khayat, K. H., Underwater repair of concrete damaged by abrasion–erosion, Ph.D. dissertation, University of California at Berkeley, 1989, 350 pages.

24. McLeish, A., A review of underwater concreting, *Proceedings, International Conference on Concrete in the Marine Environment,* The Concrete Society, London, 1986.

25. Browne, R. D., Broomfield, J. P., McAnoy, R., McLeish, A. & Robery, P. C., Diagnosis and repair of marine structures—towards a unified approach, *Proceedings, International Conference on Concrete in the Marine Environment,* The Concrete Society, London, 1986, pp. 321–32.

26. Sawada, E., Repair method for salt-damaged reinforced concrete structures, *Concrete International,* **12**(3) (1990), 37–41.

27. Coote, A. T., McKenzie, S. G. & Treadaway, K. W. J., Repair to reinforced concrete in marine structures: assessment of a method for studying and evaluating repair systems, *Concrete International,* **12**(3) (1990), 333–48.

Chapter 9

The Future of Concrete in Marine Construction

As described in Chapter 1, the saga of concrete in the marine environment is not over; in fact; it may be entering into a new era of intense activity. Oceans represent an inexhaustible source of water, food, energy and minerals which are vital to the sustenance and development of human society. Therefore, with the increasing shortages of materials on land, man's search for additional resources is now focused on coastal and offshore areas. This is what Gerwick[1] has to say on the subject:

> The last three decades have seen amazing advances into the sea, changing our concepts and opening a new frontier in the development of resources for use by society. Its vastness, its profundity hold us in awe. Only recently have humans begun to think seriously about work in the deep sea, in iceberg-prone areas, and in the Arctic . . . . In the late 1970s, engineers stated that it would be impossible to design structures capable of resisting iceberg impact. Later, it was speculated that the building of structures in the Arctic will be possible only in relatively shallow depths. Next, there was a question whether a structure could ever be built to resist multiyear floes. Now that all three of the psychological barriers for construction in the Arctic have been crossed, how much farther will the construction activity in oceans proceed in the future?

Beginning with a water depth of 70 m (Ekofisk 1, 1973), most of the concrete offshore drilling platforms already in place are at water depths ranging from 100 to 140 m. The largest concrete platform (244 000 m³ concrete), Gullfaks C, is installed at 216 m water depth. The discovery of oil and gas in deeper waters (250 to 400 m) of the North Sea has brought forth new concepts for the design of concrete platforms. According to Hoff,[2] as these structures become larger to accommodate the greater water depths and generally poor subsoil conditions that are associated with the deeper waters of the North Sea, their ability to float from the construction site to the point of installation on the sea floor many hundreds of kilometers offshore may be impaired unless light-weight aggregate concrete is used in making the structural components, at least in the upper regions of the structure. By reducing the concrete density while still maintaining high strength, tradeoffs between buoyancy and displacement can be achieved that will allow the structure to float and be towed with an adequate margin of safety.

Once in place, however, the lightweight aggregate concrete structure must meet the strength, ductility, and fatigue resistance requirements that are highly demanding. Recently, this has led to considerable research and development activity in Europe, Japan, and North America on high-strength lightweight aggregate concrete. For instance, Malhotra[3] reported that lightweight aggregate concrete with $2000 \, kg/m^3$ density and 70 MPa compressive strength could be commercially produced using a high-quality lightweight aggregate and a superplasticizer. Detailed reports on composition and behavior of high-strength lightweight aggregate concrete have resulted from other studies in Norway,[4] Canada,[5] and the United States.[6-8]

It is interesting to point out that the revised version of Norwegian Standard 3473, published in November 1989, contains design rules allowing the use of 85 MPa (Grade LC 85) light-weight aggregate concrete. Although laboratory samples of up to 100 MPa lightweight aggregate concrete are reported, with large-scale construction the practical upper limit with materials available today seems to be 70 MPa. Increasing utilization of this type of concrete for offshore marine structures in the future is predicted. According to an ACI Committee 357 Report, two lightweight concrete structures, the CIDS (Concrete Island Drilling Systems, see Chapter 1) and the Tarsuit Island Caissons, have already been deployed in the Arctic, and both have performed satisfactorily.

Also, with high-strength lightweight aggregate concrete, floating semisubs are being developed for oil exploration and production. Compared to steel semisubs, concrete semisubs are capable of carrying large topside loads and are less expensive to build, as is the case with concrete platforms. Floating bridges are another area of future application for lightweight concrete. Already under construction is a 914 m long floating bridge, connecting the city of Kristiansund to an island on the west coast of Norway. It consists of a steel frame supported by seven lightweight concrete pontoons.

Increased confidence in the satisfactory performance of high-strength normal-weight concrete has ushered in an era of unprecedented concrete construction in the marine environment. A superplatform for the 300 m deep Troll field in Norway is being designed for construction using a tension leg platform (TLP) concept developed by the Norwegian Contractors. The platform will be anchored to a foundation in the seabed by tension legs. With the commercial advent of very high-strength concrete (70 to 100 MPa), slim monotowers are now possible. The Draugen platform, which will be installed at a water depth of 251 m in the North Sea, in spite of the record height of 285 m of the concrete substructure, will use only 83 000 m³ concrete owing to the slender monotower (one column) design. This concrete will conform to Grade C 70, i.e. 70 MPa compressive strength. Sleipner A Condeep platform, which is under construction and is being designed for a service life of 50 years, will use 65 MPa concrete. The concrete platform for the Barge field, which is at present being designed, will have an 80 MPa concrete.

Prestressed superspan concrete bridges, such as the Norddalsfjorden Bridge and the Helgeland Bridge, mentioned in Chapter 1, are being constructed to cross the open seas. Bridges across the Straits of Messina, Gibraltar, and Bering are still in the conceptual stages. Given the political and economic significance of these and other similar projects, they will be built in the future when we move away from the cold-war era to a new era of global peace and understanding.

For fjord and bay-crossing projects (floating bridges, submerged tunnels, etc.) even stricter requirements on concrete quality than with offshore platforms are being contemplated. The use of exceptionally low water/cement ratios (less than

0·3) in superplasticized concrete containing silica fume (8–10% by weight of cement) and very hard particles of aggregates can result in concrete mixtures with more than 100 MPa compressive strength and high wear-resistant characteristics. Europe's first floating bridge, designed as a continuous curved pontoon 1230 m long, with a cable-stayed bridge opening for boat traffic at one end, is to be constructed near Bergen on the west coast of Norway. Originally designed with ordinary concrete, it has been redesigned using high-strength concrete to reduce its dimensions to the satisfaction of fishing interests.

Floating offshore airports in the open sea, floating industrial facilities plants for production of power and oil-processing, and floating waste-treatment facilities remain a dream of the future. However, Gerwick[1] is confident that, owing to its durability and cost-effectiveness, concrete will be ready to meet the challenge when these dreams are translated from the conceptual to the planning state. Also, such spectacular projects should in no way obscure the vast number of coastal ports, bridges, foundations and sea walls in concrete, which will be essential to meet the ongoing needs of society.

Lastly, oceans are the ultimate sink for all surficial matter. Land and surface water pollution eventually pollutes the oceans. Thus land and surface water pollution adds to the ocean pollution caused by direct dumping of untreated sewage, chemical, and nuclear wastes. To protect sea life there is already a growing public opposition to the direct dumping of these wastes into seawater. The ecology of the oceanic and the land environments will have to be integrated if the oceans of the world are to serve man as a source of food and water. This may happen sooner than we think, and such environmental considerations will also influence the type of marine construction projects we undertake and the choice of construction materials used in the future.

REFERENCES

1. Gerwick, B. C., *Construction of Offshore Structures*, Wiley–Interscience, 1986, p. 477.

2. Hoff, G. C., High-strength, lightweight aggregate concrete—current status and future needs, *Proceedings, Second International Symposium on Utilization of High Strength Concrete*, ed. W. Hestor, ACI SP-121, 1990, pp. 619–44.
3. Malhotra, V. M., CANMET Investigations in the development of high-strength lightweight concrete, *Proceedings, First International Symposium on Utilization of High Strength Concrete, Stavanger, Norway*, 1987, pp. 15–25.
4. Holand, I., High strength concrete research programme in Norway, *Proceedings, First International Symposium on Utilization of High Strength Concrete, Stavanger, Norway*, 1987, pp. 135–46.
5. Seabrook, P. I. & Wilson, H. S., High-strength lightweight concrete for use in offshore structures, *Int. J. Cement Composites and Lightweight Concrete*, **10**(3) (1988), 183–92.
6. Developmental design and testing of high-strength lightweight concrete for marine Arctic structures, *Reports on AOGA Project No. 230*, ABAM Engineers Inc., 1986.
7. Mor, A., Fatigue of high-strength lightweight aggregate concrete under marine conditions, Doctor of Engineering Dissertation, University of California at Berkeley, 1987.
8. Hoff, G. C., High-strength lightweight concrete for the Arctic, *Proceedings, Gerwick Symposium on Durability of Concrete in the Marine Environment*, ed. P. K. Mehta, Dept. of Civil Engineering, University of California at Berkeley, 1989, pp. 9–19.

Index

Printed and bound by CPI Group (UK) Ltd, Croydon, CR0 4YY

01/11/2024

01782605-0008